# ENGINE AIRFLOW

A Practical Guide to Airflow Theory, Parts Testing, Flow Bench Testing, and Analyzing Data to Increase Performance for Any Street or Racing Engine

### Harold Bettes

**HPBooks**

# HPBooks
**Published by the Penguin Group**
**Penguin Group (USA) Inc.**
**375 Hudson Street, New York, New York 10014, USA**
Penguin Group (Canada), 90 Eglinton Avenue East, Suite 700, Toronto, Ontario M4P 2Y3, Canada
(a division of Pearson Penguin Canada Inc.)
Penguin Books Ltd., 80 Strand, London WC2R 0RL, England
Penguin Group Ireland, 25 St. Stephen's Green, Dublin 2, Ireland (a division of Penguin Books Ltd.)
Penguin Group (Australia), 250 Camberwell Road, Camberwell, Victoria 3124, Australia
(a division of Pearson Australia Group Pty. Ltd.)
Penguin Books India Pvt. Ltd., 11 Community Centre, Panchsheel Park, New Delhi—110 017, India
Penguin Group (NZ), 67 Apollo Drive, Rosedale, North Shore 0632, New Zealand
(a division of Pearson New Zealand Ltd.)
Penguin Books (South Africa) (Pty.) Ltd., 24 Sturdee Avenue, Rosebank, Johannesburg 2196, South Africa

Penguin Books Ltd., Registered Offices: 80 Strand, London WC2R 0RL, England

While the author has made every effort to provide accurate telephone numbers and Internet addresses at the time of publication, neither the publisher nor the author assumes any responsibility for errors, or for changes that occur after publication. Further, the publisher does not have any control over and does not assume any responsibility for author or third-party websites or their content.

**ENGINE AIRFLOW**

Copyright © 2010 by Harold Bettes
Cover design by Bird Studios
Front cover illustration by James Beaver
Back cover photos by Harold Bettes and Endyn
Interior photos by author unless otherwise noted

All rights reserved. No part of this book may be reproduced, scanned, or distributed in any printed or electronic form without permission. Please do not participate in or encourage piracy of copyrighted materials in violation of the author's rights. Purchase only authorized editions.
HPBooks is a trademark of Penguin Group (USA) Inc.

First edition: July 2010

ISBN: 978-1-55788-537-1

PRINTED IN THE UNITED STATES OF AMERICA

10  9  8  7

NOTICE: The information in this book is true and complete to the best of our knowledge. All recommendations on parts and procedures are made without any guarantees on the part of the author or the publisher. Tampering with, altering, modifying, or removing any emissions-control device is a violation of federal law. Author and publisher disclaim all liability incurred in connection with the use of this information. We recognize that some words, engine names, model names, and designations mentioned in this book are the property of the trademark holder and are used for identification purposes only. This is not an official publication.

# CONTENTS

**Acknowledgments** — iv
**Foreword by Jim McFarland** — iv
**Introduction** — v

**Chapter 1**
**Basic Engine Terms, Concepts & Components** — 1

**Chapter 2**
**Internal Combustion Engine Fundamentals** — 17

**Chapter 3**
**Engine Airflow Relationships** — 38

**Chapter 4**
**How Flow Benches Work** — 44

**Chapter 5**
**Flow Testing Tools, Measurements & Calculations** — 51

**Chapter 6**
**Establishing Airflow Testing Standards** — 72

**Chapter 7**
**Airflow Calibration & Measurement Standards** — 75

**Chapter 8**
**Comparing Flow Numbers** — 79

**Chapter 9**
**Flow Testing Intake & Exhaust Components** — 85

**Chapter 10**
**Flow Testing Carburetors, Fuel Injection & Manifolds** — 113

**Chapter 11**
**Selecting Camshafts with Flow Numbers** — 125

**Chapter 12**
**Graphical Analysis** — 128

**Chapter 13**
**Computer Simulation Programs** — 141

**Handy Formulae & Charts for Gearheads** — 144
**Unit Conversions & Equivalents** — 149
**Resources** — 150
**Recommended Reading** — 152

# ACKNOWLEDGMENTS

Thanks to my teachers and professors that had enough patience to teach me how to learn. I remember all the good ones, especially Dr. Dean Hill.

A special thanks to Jack May, an old pal who as a brother Quarterhorses car club member has been an unwavering friend of almost half a century. He has literally had my back on more than one occasion. Thanks, Jack.

A very special recognition is due to my editor, Michael Lutfy. Michael has been patient beyond measure, and it is greatly appreciated. His patience and encouragement have made it easier to produce these words.

# FOREWORD

If the amount of knowledge in this book could be equated to weight, it's doubtful you could pick it up. Harold Bettes has a rich and long history of providing information "worth its weight in gold," and this publication is clearly no exception to that reputation.

For the more than thirty-plus years of our lasting friendship, I have been privileged to sometimes observe and other times participate in the many technical exploits of his still-growing career. Since we both have engineering backgrounds and an intense interest in internal combustion engines, we continue to share common platforms of knowledge and curiosity about how they work and can be improved.

Over time, Harold's many technical skills have been applied in ways even he might not have expected. For example, two major corporations in the automotive specialty parts industry (MSD Ignition and a once popular dyno company) were fertilized into major growth by both his schooled and intuitive ways of building a business while focusing on the critical importance of customer satisfaction. In fact, he still recognizes the value of providing fundamental engineering support to users of products that benefit from that access. He continues to demonstrate a missionary's zeal for providing understandable and thought-provoking explanations to otherwise complex subjects. The book you are holding is an example of that passion.

He has been a student and teacher of fluid flow for almost forty years, applying his skills and interests to topics presented in this book. More important, Harold's unique ability to provide simplified explanations about complex concepts is a direct path toward applying them to practical applications. And that's key to this book. Typically in the academic community, where subjects dealing with fluid flow often require an in-depth technical background, reducing theoretical notions to practice can be problematic. Based on his experience on both sides of this equation, Harold's approach removes any masks of academia and helps you "turn the wrenches" in conducting airflow studies.

In the grand scheme of teaching, it is rare to gain access to people who have a clear vision of their objectives and the skills to take you there. Quoting Harold, "simple is elegant" and "complex is arrogant." He subscribes to the Mark Twain notion that "fact is stranger than fiction" and he has the mind-set and abilities to address the unskilled and PhD, all in the same breath or sentence.

As a Texan with a healthy respect for the fundamental Texas Ranger credo of "Ride well, shoot straight and speak the truth," you will discover this book to be an editorial representation of that objective and a reflection of the person who authored its text.

All that remains is for you to enjoy and benefit from Harold's hard work. — *Jim McFarland*

# INTRODUCTION

Airflow and engines certainly need no introduction, but I probably do. I have been fortunate enough to have lived during a major period of motorsports history while the study of airflow through engines has made leaps and bounds in discoveries and development. During this time, many of the major names in engine development have been friends, regular contacts and acquaintances. What an exciting time it has been!

I actually started to write this book at the urging of long-time friend and former HPBooks editorial director, Tom Monroe, back in 1987. Although dedication to my career as an airflow, application and sales engineer, to being a good husband and father, pushed the writing of the book to the side, I can honestly say that it now benefits from over 40 years of knowledge and wisdom. It is a much better book than the one I would have written in 1987!

Most of the numerical unit references in this book will be in English Engineer's units (US measurements) for simplicity and continuity.

I've also chosen to use the mathematical expression for two flow testing standards of reference: inches of water and inches of mercury are shown as "in.$H_2O$" and "in.Hg" respectively.

As you use this handbook, I recommend you have a calculator and notebook nearby. The math is simple and can even be a lot of fun, but you will need to learn to be comfortable in manipulating various numbers to help you apply what you learn here. One thing that I have always tried to push over the years is that numbers only have meaning if you know where they come from. I have done my best to show where the numbers come from so that you'll truly learn how to interpret them. Learning how to use those numbers is a very handy skill you will use throughout your career as an auto enthusiast or engineer.

## Math Formatting

A brief explanation on how I've formatted the arithmetic in this book: Any formula or equation written in the following format $ABCD = E$ indicates that each element is to be multiplied times the next to get the final results. However, for the sake of clarity, I have also used the form $A \times B \times C \times D = E$ so that you can follow each equation very easily as the x's denote multiplication. Note that there are other signs such as $\sqrt{\phantom{x}}$, which means the square root. The sign written as $\sqrt[3]{\phantom{x}}$ means the cube root. Any use of exponents is identified as $(A)^2$ or $(B)^3$ or $(C)^4$, which means either to square, cube or the fourth power of the amount in parenthesis. Some equations have constants represented that are not normally explained. Constants are numbers that are used to make the units come out correctly or used to simplify the overall calculation process.

Each of the equations used in this book is written so that it can be entered into most simple calculators following a standard algebraic process. Math is a learned skill and doesn't come naturally to a lot of people, but it can be a very useful tool, so keep working at it and it will become easier as you learn where the numbers come from. I'll do my best to walk you through the process.

What this book is not will become pretty obvious. It is not really a textbook although it could be used to reinforce more in-depth technical material on internal combustion engines. It is also not a book that will explain exactly how to port cylinder heads and manifolds. However, it will describe how to collect or use dependable data and how to analyze it by applying sound and logical principles. The information applied correctly will help you to modify heads and manifolds and other such components with a better understanding of some goals and objectives that are targeted to help you make more power. It might even help you to affect some changes in components so that they will be able to burn fuel more efficiently. It is assumed that you have a basic working knowledge of reciprocating internal combustion engines. Although some basics of engine operation are explained, an in-depth study of the individual components is not included.

I'm assuming that the real reason you picked up this book was to learn something about airflow through engines. I hope that by the time you're done reading it, you'll have improved your knowledge to make a difference in your engine projects. Enjoy the process and never stop learning as you go through this life. As the old saying goes, "Busy hands are happy hands." Stay happy.

—*Harold Bettes II*

# DEDICATION

For my mom and dad, who taught me to work hard, to enjoy learning, and the importance of family, freedom and faith in the Almighty.

For William E. (Bill) Hofer, a trusted mentor, former WWII POW and fellow gearhead. After my own military service, he and my dad helped me to heal. Although Dad and Bill have passed on, I think they would have really enjoyed this book. They were true representatives of one of the greatest generations of this country.

A very special dedication and thanks to my dearest wife, Paula, who is and has always been the wind beneath my wings for the past 37 years and the devoted mother of our three wonderful and amazing children. I am also very lucky that she is an accomplished librarian and such a great research assistant!

*Inside an engine's intake, combustion and exhaust tract, a wind blows. This wind is the essence of the engine's performance potential. Those who seek to tame, to master this wind, the future will bring unimaginable reward. Tame the wind...*
—Steve Boszo, Airflow Researcher and Gearhead

# Chapter 1
# Basic Engine Terms, Concepts & Components

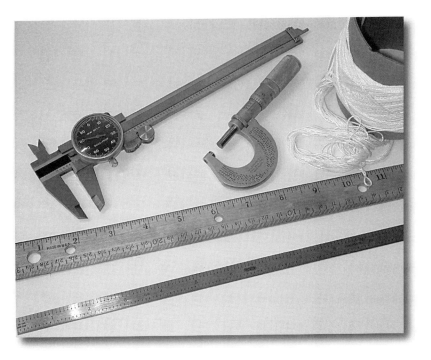

*Truth is stranger than fiction, but it is because Fiction is obliged to stick to possibilities; Truth isn't.*
—Mark Twain

**It doesn't make much difference what you use to measure components, just so long as they are quality measurement devices capable of the greatest accuracy possible. You need to learn the difference between accuracy and repeatability.**

While the terms and concepts in this chapter are by no means a complete list, they must be understood if you are to improve your knowledge of engine airflow and performance in general. Many of them you might already know, but at the very least they will be a good review, and you might learn a thing or two. They are also in no particular order of importance.

## Measuring Tools

Being involved with internal combustion engines means you'll have to become very good with a specific branch of science called metrology (the science of measurement). You can learn to measure with a multiple of devices. You can measure a length or a diameter with a stick, string, a dial caliper, or a known and calibrated micrometer and all will give you an answer, but one is much more preferable when the need to compare data with another site or have an acceptable level of certainty (accuracy) and repeatability (the ability to repeat or duplicate the measurement). If we were all talking about a crankpin that measured 2.000" diameter at one place (where it was machined and finished on a crank grinder) and someone else measures it at a diameter of 2.003" which one is correct? Unless one uses a standard of measurement that is an agreed upon standard, it is just so much guesswork. It never ends when one is constantly searching for the details that make a difference. Learn how to measure. It is a necessary task when dealing with engines and their components.

So, what should you look at if you were considering

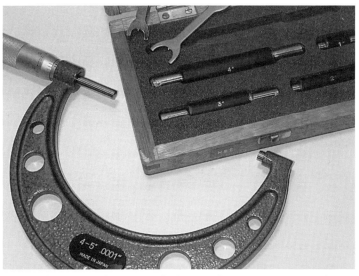

**Choose Starret or Mitutoyo or similar quality measurement tools. They will give you a lifetime of service if you take care of them properly. Normal accuracy for these types of inspection tools is +/- 1/10 of one thousandth of an inch (0.0001"). Micrometers are a necessary tool if you are going to work with engines.**

purchasing some measurement tools? Roger "Riceman" Lee, a calibration and measurement expert, says, "I would recommend quality brands like Starrett or Mitutoyo. Calibration for inspection tools should follow ANSI Z540-1 procedures. The calibration company should be ISO 9001-2000 and ISO 17025 certified or compliant."

# Engine Airflow

The very popular dial caliper is a measurement tool for rapid reference but it is normally only precise to +/- 0.001" if that much. For precise measurements, use micrometers.

A modified set of calipers can be used like this to verify and record the short side to the deck on some heads or other tasks. Inside measurement "horns" were ground off so they would fit. Photo courtesy Meaux Racing Heads.

Any race car or other vehicle is pushed around by the power that its engine produces. Timing slips prove or disprove what sometimes are outlandish claims. "The older I get the faster I was" might mean more to some folks than others. This stuff is not magic.

It is very much the same when one must compare notes on terminology and measurements and how to use that data and turn it into useful information. As these chapters unfold, you will get used to new terminology that you will need to be comfortable with in order to grow in your knowledge of engines. Learn this stuff not for the buzzwords, but to know what it all means. Learn how it works and how to apply it, too.

The internal combustion piston engine is a complex system that is made up of many parts and components that must work together if the machine is to produce sufficient levels of horsepower to propel a vehicle. The necessity of the components to work together is based upon many interactive relationships, so this book will help you to evaluate these parts combinations, since there is no magical horsepower answer. The key to component selection is make sure it is complementary to the overall engine package; the more complementary these components are, the more efficient the engine is at producing power. It is quite easy to build a fuel hog of an engine that puts out an amazing amount of horsepower. The trick is to put out the same power by burning much less fuel.

## By the Numbers

Nothing gets a gearhead's attention like the sound and smell of a internal combustion piston engine being started up for the first time. The excitement can be just as intense whether starting a restored hundred-year-old engine, or a current NASCAR racer, Top Fuel dragster or '32 Ford street rod. But what every true gearhead wants to know, regardless of the engine type, is how much horsepower makes. Even though a bystander might not understand the specific details of engines or horsepower, there is interest in the numbers, and bigger numbers are always more impressive than smaller ones.

The engine that we all get so interested in is a system of both some simple and complex components that must work together in order to produce power. The more complimentary their interactions, the better the level of power for a given amount of fuel and air consumed. It is the specific chain of events and the interaction of each that allows some engines to have something that others just don't have—character and the smooth production of power

There are many myths surrounding the internal combustion (IC) piston engine and how to increase horsepower, but it is not magic or any great secret. Some people try to create a shroud of mystery behind airflow, because it can't really be seen. But once you learn the basics and fundamental theories, it really won't seem so complex.

To put it in the most basic terms, it is the burning of fuel that makes power, which requires an adequate amount of air to accomplish. The more air, the more fuel that can be burned more completely in the combustion chamber, and therefore the more power that can be produced. That is true whether the fuel is gasoline, diesel or kerosene, methanol, ethanol, E10, E85, butane, propane, hydrogen, natural gas, nitropropane, nitromethane or various mixtures thereof. It even applies to biodiesel or the grease that your French fries were cooked in.

# Basic Engine Terms, Concepts & Components

This huge engine is one great colossal stack of metal and thought process. Each cylinder makes 7,780 horsepower. The photo shown is only a 12 cylinder version of the largest engine which is a shipboard unit that has 14 cylinders. This awesome thing is 44 feet high!

Scale model engines like these are truly works of art as much as science. On the left is a little Offenhauser 170 while the one on the right is an injected Ford with Ardun heads. The builder is Ron Bement of Denver. They have exactly the same characteristics as their full-sized counterparts. You should hear these things run.

*The Largest Engine*—At the time of this writing, the largest and most powerful engine in the world is a 14-cylinder inline, turbocharged, two-stroke diesel engine that burns bunker fuel oil (heavy oil) and it was built to power mega sized container ships. The engine weighs 2,300 tons (4,600,000 pounds) and is 89 feet long and 44 feet high. The bore is approximately 38" and the stroke is approximately 98". The displacement of this huge engine is 1,556,002 cubic inches! The engine produces 108,920 horsepower at 102 rpm. The torque produced is a phenomenal 5,608,312 lbs-ft. at 102 rpm. The fuel consumption rate is 1660 gallons of heavy oil per hour. That means in a 24-hour period at sea and making maximum rated power, this engine burns nearly 40,000 gallons of fuel. The crankshaft has main and rod journals that are 38" in diameter (equal to the bore diameter). Smaller engine versions are available in 6 and 8 cylinder inline configurations that also produce 7,780 horsepower per cylinder!

*Smallest Engines*—At the time of this writing, the smallest engines in the world were very diverse in design. The smallest displacement fuel engine is a glow plug two stroke that displaces a miniscule 0.010 cubic inches. The world's smallest four-stroke engine is a tiny four-cylinder spark ignition engine that displaces only 0.061 cubic inches with a bore of 0.250" and a stroke of 0.3125" (that is only 0.0153 cubic inches per cylinder). The little engine is only 1.9" long and 1.19" high and weighs only a few ounces. The camshaft is about the size of a fat kitchen matchstick. There are also tiny Wankel rotary-type engines that are about the size of a penny and supposedly gas turbines that are just about atom sized units.

The future holds many unknown, exciting developments in engine technology, but one thing is for certain; if an engine breathes air and burns fuel then the study of how it handles air will always be a challenge.

## Basic Power Theory

*Horsepower*—A Scottish engineer by the name of James Watt (1736–1819) coined the term horsepower in about 1780. Watt needed a term for power that would help him market his modified steam engine. It made sense to call the power level "horsepower," because at the time that is the other "engine" he was competing with—a horse. Because pulling and lifting loads had to be related and equated to rotational energy (the steam engine had a rotating shaft), Watt discovered that a good average of work done by a harnessed horse was moving 33,000 lbs. (rounded off number) 1 foot in one minute. It took something like a huge Clydesdale (just like those that pull the Budweiser beer wagons) to do the work for any period of time that Watt chose as an example of the standard of one horsepower. No pony power was considered.

The engineering reference for work done is a force multiplied times a distance. So, the weight of 33,000 lbs. moved a distance of 1 foot is 33,000 ft-lbs. of *work*. Note that this is a reference for work, not *torque*. That will be explained shortly. The engineering reference for power is work divided by time or work per unit of time, which is equal to power. When we call it horsepower, the reference is 33,000 ft-lbs. of work per minute (ft-lbs/min) or 550 ft-lbs/sec. Try and remember these numbers as they are the basis for others as you learn more about where the numbers came from.

If you wanted to relate the above description to a machine that had a rotating crankshaft, then one

# ENGINE AIRFLOW

To get on the podium takes lots of preparation and attention to the details. Understanding the application of the details is part of the road to success. The teams these drivers represent all had done their homework about power and reliability. Photo courtesy Virkler and Bartlett.

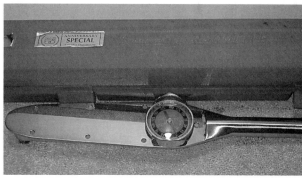

What you see here is obviously a torque wrench and they are typically mislabeled. They are marked (scaled) as ft-lbs., but the proper method would be pounds-feet (lbs-ft.). The text will explain the difference.

could take the rotation of the crankshaft and say that was equal to a rotational distance of $2\pi$ (which is 360°), because it takes $2\pi$ of those to equal one complete revolution. If you divide the 33,000 above by $2\pi$, the answer is 5252.1 and if you round that off, the constant 5252 is revealed. Now you have a piece of history that can come in handy if anyone wants to play trivial information games. Or perhaps it might provide you with some cold liquid refreshment if it was part of a bet. Either way is OK as long as you remember it and where it came from. So, by definition:

**HP = (T x rpm) ÷ 5252**

With some gentle algebraic manipulation, the other elements of the equation can be shown so that the following can be applied as required.

**T = (5252 x hp) ÷ rpm**
**rpm = (hp x 5252) ÷ T**
**where hp = horsepower, T = torque, rpm = revolutions per minute and the constant 5252 is from the previous explanation above.**

You need to study the above enough so that you have a very good understanding of the relationships of torque, power and rpm. It is on this basis of understanding that your attention to detail will grow as you study engines and their applications. At first it doesn't seem like much, but the basics are the building blocks for all sorts of dreams and aspirations. Commit these relationships and equations to memory. It helps to use them when you can in calculations.

***Torque vs. Horsepower***—The phrase "I would rather have more torque than horsepower..." is often-heard and misguided. If you learn the difference between the two, you might be able to win a cold one or two during a heated argument with other gearheads on the true definition of each.

By definition, *torque* is a twisting force multiplied by a lever's length. This torque is also called a *moment*. The engine generates torque as a result of a force upon the piston top, which pushes the connecting rod down and produces the twisting action of the rotating crankshaft. In engineering terms, *instantaneous torque* is what happens at any given moment (pun intended) during the expansion cycle. That force, pushing on top of the piston (sometimes referred to as *IMEP* or *indicated mean effective pressure*) is a result of expanding burning gases. The internal friction is referred to as *FMEP* or *friction mean effective pressure*. The leftover effort is called *BMEP* or *brake mean effective pressure*. This is all explained in greater detail in Chapter 2, page 22.

By the way, torque is typically expressed in pounds-feet (lbs-ft.), not foot-pounds (ft-lbs.). Yes, it does make a difference. Yeah, torque wrenches are often incorrectly labeled if they have ft-lbs. on their scales and it is done all the time.

*Horsepower* is a calculated number that has much more emotion tied to it than is really necessary in order to evaluate or learn about the subject. As mentioned, James Watt used the term to relate the capabilities of his steam engine to the most popular mode of power of that day in 1780. However the first reference to *brake* horsepower was not until 1821. Much changed during those years, but the need to understand the terms and applications has pretty much remained constant when studying about reciprocating internal combustion engines.

As mentioned above, hp = (torque x rpm) ÷ 5252. So, note that by this definition, torque and horsepower are the same value at 5252 rpm. In the

## Basic Engine Terms, Concepts & Components

When it is time to line your car up to race, often the difference in winning and losing is a firm understanding of the difference in horsepower and torque. In order to pick up just a few miles per hour of speed takes more power that is applied to the track surface.

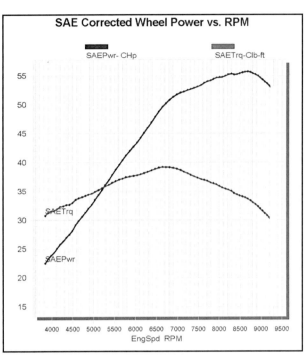

This graphic is for a twin-cylinder Honda motorcycle after cylinder head rework. Graph is from a chassis dyno test and the net results picked up the power from a "before" of 40 hp to an "after" of 58 hp. Impressive gain. SAE Corrected Power is at the rear wheel tire patch.

horsepower formula, the rpm is entered as a time function. Note that because torque and rpm have equal positions in the formula (each is a multiplier), it takes both to produce a horsepower number.

Horsepower is what pushes, pulls, shoves or drags our vehicles (boats and aircraft as well) through this friction-filled world. If we want to go faster, it takes more horsepower. If we want to go slower, it takes less horsepower. How much more power does it take to go faster? It takes a lot more than you would think. In order to go twice as fast, it takes 8 times more power. In short, if you have a vehicle that will go 100 mph, then it will take 8 times the amount of horsepower that produced the 100 mph speed to go 200 mph! In real numbers:

If it requires a power level of 90 horsepower for your vehicle to go 100 mph (146.7 ft/sec), then it will take at least 720 horsepower to reach 200 mph (293.3 ft/sec). Think about that for a moment and let it sink in. It is substantially easier to talk about it than it is to do. That is why so many people will try and tell an exciting story about how fast their car might be or how fast they've driven before. Yep, everybody can go 150 mph (220 ft/sec) until you get to actually time the vehicle over a known distance and they will almost always come up very short of the claimed goal. Speed and power are much easier to say than to accomplish. Learn to understand the differences.

What about the folks that still would say that they would rather have torque than horsepower? Tell them that at least one option is to leave their vehicle in low gear because that is typically the greatest torque multiplication that is going to be available! In low gear, the torque is multiplied, but not the horsepower. The fastest and quickest vehicles have more average horsepower. Peak power doesn't make it by itself nor does peak torque. It is the combination with the greatest area under the whole power curve that has the advantage.

Hopefully that will become more obvious as we continue into the nuances of engines and how they function.

Some real-world applications and evaluations of torque and horsepower:

Torque specifications: Tighten lug nuts to 70 lbs-ft. or tighten head bolts to 75-80 lbs-ft. or torque manifold fasteners to 25–30 lbs-ft.

Typical engine torque specifications: 500 lbs-ft. @ 5000 rpm (note it has an rpm reference). Horsepower in this example: hp = (500 x 5000) ÷ 5252 = 476.01 hp.

Remember that you cannot have horsepower without torque, but you can have torque without horsepower! Think torque wrench.

***Variations of Horsepower & Torque***—Horsepower and torque have many variations when it comes to engines. Terms such as *Fhp (friction horsepower)*, *CBhp (corrected brake horsepower)*, *Ohp (observed horsepower)*, *Ihp (indicated horsepower)*, *FT (friction torque)*, *CBT (corrected brake torque)*, *OT (observed torque)*, *IT (indicated torque)* are some of the ones we'll be discussing. Note that there is a difference in the various methods of correcting horsepower numbers even though they are still called horsepower. The many exacting details of

# ENGINE AIRFLOW

engines cannot be properly presented in such a short book, but we can discuss many of the more interesting engine relationships when we have the opportunity.

Also to be discussed is how to improve torque and horsepower production with the study and application of engine airflow, which is what this book is about anyway. If improved performance and horsepower is the intended target, one needs to learn how to get to that goal by understanding the details that make up the complete package of components. It involves the skillful selection of those components, not one single super magic trick although some folks might believe in that particular myth. You can study in order to gain the required knowledge to overcome the unknown.

## Fluid

The term *fluid* is used to describe all gases and liquids. A more classic definition is an amorphous (shapeless) substance whose molecules pass each other without hindrance—the substance follows freely, seeking the limits of the container. The study of fluids is the basis of fluid mechanics, which involves both fluid statics and fluid dynamics.

Obviously, if the fluid is in motion or not defines which method of study. That leads us to the fact that airflow through engines is a study of fluids. Fluid motion of the air through the engine is motion that ultimately allows the production of horsepower to occur. If there is no fluid motion, or *airflow*, then there is no power (at least on this planet). It is very easy to remember that air moves by pressure differentials. If the pressure in the intake tract is the same as the pressure on the other side of the inlet valve, no flow will result. So studying various pressure differences and how that happens will enhance your understanding of fluid flow.

## Top Dead Center (TDC)

This refers to the location of the piston versus the stroke or offset of the crankshaft. *Top dead center* is when the piston is at the location at the top of the stroke where it is momentarily motionless. An easy way to find TDC is to use a positive stop placed in a spark plug hole. Or, you can use a dial indicator if the heads are off. Then you need to attach a degree wheel to the crank snout. On cylinder number 1 of your engine, using the stop or the indicator, turn the engine in the clockwise direction until the piston is about 0.200" or so below the topmost movement. When that position is referenced on the degree wheel, turn the crankshaft counterclockwise until the piston is back to the same reference below the uppermost travel (0.200" in this example) and

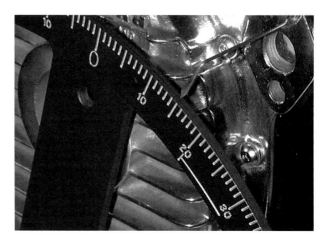

TDC is a super important item to know about your engine and this shows part of the process in establishing that reference which in this case is precisely at 15 degrees before TDC (BTDC). It is not uncommon for factory TDC marks to be way off.

record the position of the degree wheel. One-half of the distance between the two reference points on the degree wheel will be mechanical TDC (you will have to remove the positive stop to get to TDC). At that point you can align the degree wheel to its marking of TDC and your pointer. This is also a great time to mark (or verify the mark) on the harmonic balancer or the crank hub (see below). While you have the degree wheel in place, you can use it to degree in the camshaft and also to verify the position of other cylinders relative to cylinder number one. The all-important reference to TDC will stay with that particular engine all its life. TDC should be verified on all engines that you might work with so that you do not make incorrect assumptions. This simple yet detailed activity can add power to your bucket of bolts.

*Checking Harmonic Balancer*—Accurately marking off the harmonic balancer of an engine is a pretty simple exercise. First measure either its diameter with a dial caliper or use a flexible tape measure and measure its circumference. The circumference can be calculated by application of the equation:

$C = \pi D$
where C = circumference in inches, $\pi$ = 3.1416 and D = diameter in inches

Once you have the established the circumference in inches, simply divide the circumference by 360 and you have the answer in inches per degree. The reciprocal of that number (using the 1/x key on your calculator) is a number for degrees per inch. I prefer the degrees per inch answer so that you can

## Basic Engine Terms, Concepts & Components

then mark off the balancer at spaces that represent what you want marked on the balancer for timing references such as 10, 15, 20, 25, 30, 40 degrees BTDC (before top dead center). One final note here is that once you have established a true TDC location, the marks for ignition timing will be to the right of the TDC mark as you face the balancer for clockwise rotation engines.

### Bottom Dead Center (BDC)

This refers to the location of the piston versus the stroke of the crankshaft. BDC is where the piston is at the very bottom of the stroke and the piston movement is momentarily motionless. This point will be exactly 180 degrees opposite TDC on your degree wheel, assuming that the piston pin bore is not offset. However, the pin offset has a very minor effect.

### Cylinder Bore Diameter

The diameter of the cylinder bore is normally measured in inches. It is common to give the bore dimension to at least three decimal places, as in 4.005". A quick measurement can be done using a dial caliper (normal accuracy of most dial calipers is only +/- 0.001"), but the bore dimension is best verified by the use of either a telescoping snap gauge or preferably an inside micrometer or a dial bore gauge. The bore dimension should be recorded for all the individual cylinders. The normal accuracy of inside micrometers should be at least +/- 0.0002" and the normal accuracy of a dial bore gauge should be at least +/-.0005. Taking several measurements and averaging them would also help improve the final dimension references and tolerances. Do not forget that for any measurement to be accurate, it must be compared against an approved Standard from which a calibration can be done. A repeatable measurement is also important on any engine component. Can you measure the bore with a wooden ruler or a piece of string? Sure you can, but the accuracy of the measurement would be lacking compared to other instruments.

### Stroke

*Stroke* is a reference to the crankshaft and the motion that it imparts upon the pistons via the connecting rods. The stroke is the amount of movement that the piston travels from TDC to BDC. A 3.00" stroke has an offset from the center of the main journals of 1.50" and because it moves up and down, the stroke length is 1.50" x 2 = 3.00". The offset as described is called the *throw* or *radius* of the stroke. In short, a crankshaft with a 3.00" stroke has a stroke radius (throw) of 1.50".

**Locating TDC is an important chore that must be done by some simple methods explained in the text. The photo on the left is a positive stop and the photo on the right is using a very large degree wheel. The maximum upward location of the piston is established and a degree wheel is used to make it more exact. Read the text carefully.**

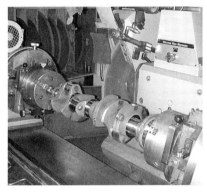 

**The crankshaft is shown in a crank grinding machine at left that ensures the sizes of the journals are correct and that the stroke is also correct. Serious machine work early on makes assembly much easier. Learning how to make proper crankshaft stroke measurements is a very handy skill for any engine builder.**

**The large holes in the block are called bores and must be precisely machined in preparation for making new homes for the pistons. This photo was taken during the honing process that ensures the holes are round and straight. This photo was taken just after the special plates were removed from the top of the block.**

There are special measurement instruments that are comparatively inexpensive to measure the stroke of a crankshaft before it is installed in the engine block. Oddly enough, not a lot of folks understand this most simple of measurements. It is literally how far the piston travels from TDC to BDC. So for a 4.00" stroke, it moves up 4.00" from BDC to TDC and the offset of the crankshaft throw is 2.00". Make a sketch and think about it until you understand it. Understanding the mechanism will help everything else fall into place.

## Displacement

*Displacement* is the swept volume of the engine. *Swept volume* means that the crankshaft rotates through two full revolutions on a four-cycle engine. An easy way to calculate the displacement of an engine is using the following equation:

$CID = B^2 \times S \times 0.7854 \times N_{cyl}$

where CID = cubic inches displacement, B = bore dimension in inches, S = stroke dimension in inches, $N_{cyl}$ = number of cylinders. The number 0.7854 comes from $\pi \div 4$.

Several different types of pistons are shown in this photo. The two in the center of the group are for mega dollar Formula 1 engines and if you look closely you will see that they only use two ring grooves instead of the more normal three grooves. The piston is just a convenient platform for the piston rings.

Commit this one to memory. There is an old saying that goes "there is no replacement for displacement" and although that is novel, it is not necessarily the truth.

In general, torque is proportional to engine displacement. So, that is the premise of the old racer's tale of "there is no substitute for cubic inches." There are specific advantages to displacement when limited by racing rules and regulations, but the bore and stroke combination chosen can be a skillful decision. If you wish to state the displacement of the engine in liters, you must multiply the cubic inches by 0.01638706 to obtain liters. There are 61.02374 cubic inches per one liter. There are 16.38706 ccs's per one cubic inch and that gives us the power to be quite international if we wish. So a 500 cubic-inch Pro Stock engine is also 8.193532 liters or 8,193.53 cubic centimeters.

## Detonation

*Detonation* is a term that refers to a condition of uncontrolled combustion, which is very destructive to engine parts. Sometimes it can be heard above the other engine noises. Sometimes it is referred to as "spark knock" or the even more descriptive expression "death rattle." Detonation is generally a result of too much ignition timing or a compression ratio too high for the fuel being used. Do not take this condition lightly.

## Piston

The piston provides a platform for the piston rings, and its geometry (at TDC) provides part of the combustion chamber. It is usually made of aluminum alloy but in the past, various materials such as magnesium, aluminum-beryllium alloy, cast iron, titanium and other exotics have been used with varying success. The topside of the piston sees higher temperatures while the underside sees much less temperature. It is common in endurance engine applications that the pistons get annealed (material gets softer) and must be replaced after some time. The bottom side of pistons is sometimes exposed to oil sprayers or "squirters" that help to control crown temperatures of the piston material and also lubricate the piston pin and pin bore for better longevity. The pin bore can be placed in either the geometric center of the piston or it can be offset to the right or left any amount. The piston manufacturer and your engine builder will help you with that decision.

The pin can be retained in the small end of the connecting rod by either interference press fit or it can be used as a floating item and retained in the piston by circular clips of various designs. Retention of the pin is a critical item in the long list of critical items in an engine.

*Piston Area*—In general, power is proportional to total piston area. Calculate the piston area by:

**Piston Area ($A_p$) = $B^2 \times 0.7854 \times N_{cyl}$**
where $A_p$ = area of the piston in square inches,
B = bore dimension in inches,
$N_{cyl}$ = number of cylinders

Try to remember that power is proportional to piston area.

## Compression Height (CH)

*Compression height* refers to the distance on the piston from the center of the piston pin bore to the top of the piston. The top refers to the flat portion of the piston that will be closest to the top of the block when the piston is at TDC. This measurement is normally given in inches and is

# Basic Engine Terms, Concepts & Components

This is a profile of the piston. Note the distance from the center of the pin bore to the top of the piston flat is the compression height. Look closely at this photo for reference as the text explains more about pistons.

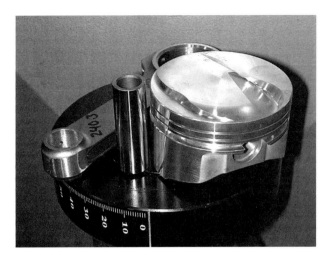

There are a lot of things to look at in this photo. The pin end of the connecting rod and the piston pin and the piston are obvious. The raised portion of the piston fills the combustion chamber when the piston is at TDC and increases the CR. This piston is for 10.2:1 with the correct combustion chamber volume.

sometimes called *pin height*. The compression height of the piston is a critical measurement in design so that the pin has room to fit and allow enough room for an appropriate ring package. Any material above the flat portion of the piston is called a dome and has a positive effect on the compression ratio of the engine (making the number higher). Conversely if the piston has a negative effect (making the compression ratio lower) by having a reverse dome or dish relief below the flat of the piston, that must be considered when calculating the compression ratio. For more on compression ratio, see page 11.

## Piston Pin

The piston pin connects the piston to the small end bore at the top of the connecting rod. It is also sometimes referred to as a wrist pin or a gudgeon pin. The piston pin is subject to considerable bending and stresses at both TDC and BDC.

Although it is a trend to choose piston pins that are lightweight in order to decrease reciprocating mass, one must be careful to consider the strength required for this critical component. Sometimes a more skillful choice of these components is to select for larger diameter, shorter length, and less wall thickness than other designs. Bending piston pins are commonly the cause of many supposed piston failures. If the pin bends and reduces the oil clearance between the pin bore and the pin, the piston material galls and seizes. The destructive result will pull the bottom out of the pin bores from the piston thus failing the piston. The small end of the connecting rod and the pin bores in the piston are prone to gall if not properly lubricated. The piston pins normally get oil from splash only, but some have pressure oiling from drilled connecting rods or have piston oil squirters installed in the engine to also cool the bottom of the piston crown.

## Piston Rings

The piston rings provide cylinder sealing and oil control and are fitted to the pistons.

If the rings do not seal well, the result is called uncontrolled blowby and is literally the leakage past the rings into the crankcase. Poor ring seal is often the cause for producing blue smoke in the exhaust (junk valve guides or seals can cause the same thing). Ring designs are many and each has specific applications. The most common configuration for a ring package is a top (compression) ring, a second (scraper) ring, and a third (oil) ring. Some diesels have up to four rings and some special engines only use two rings (compression and oil) with good success. Engineering studies have shown about 70% of an engine's internal friction comes from the piston and ring package. Rings come in all sorts of shapes and sizes.

Compression rings are typically 0.039" (1mm) to 0.078" in thickness. The common trend in performance applications is to use rings that are

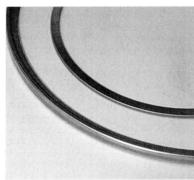

Rings are an absolute necessity in order to seal against the bore surface to capture the pressure in the cylinder bore. There are many different types of piston rings. The photo on the right shows a special step type ring and a very thin ring. Materials are explained in the text. Normal applications use a top or compression ring a second which is a scraper and the lower ring is an oil ring which provides oil control.

**This is a tiny oil ring used in a tiny bore of a small V8 engine. The function and importance of the rings are the same regardless of engine size. Photo courtesy Conley Precision Engines.**

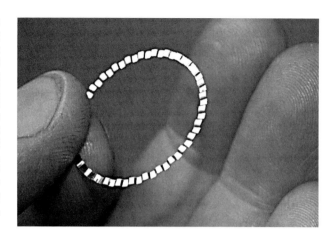

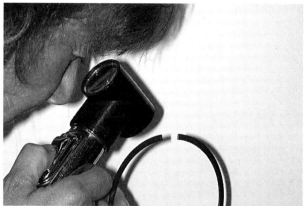

**A piston ring has many surprising functions. They become more apparent when you look at a ring through a spark plug light.**

0.039" (1 mm) to 0.062" (1.59 mm) thick. There are a few applications where rings are only 0.023" to 0.027" (0.6 and 0.7 mm) thick. Because of the greater unit loading on the smaller surface, thin rings wear at a greater rate than thicker rings. The thin rings are also more sensitive to heat because they have less mass. The rings must seal on the sides (in the piston grooves) as well as on their surfaces in contact with the cylinder walls. The required surface finish of the ring grooves dictates that particular area should never be glass beaded in cleaning process for rebuilds. It is pretty common for NASCAR engine builders to anodize the piston ring grooves to make the surface hardness handle the stress of endurance events better than the standard softer aluminum material. The aluminum oxide finish that anodizing provides is much harder than the aluminum parent material. An anodized (type III) surface is typically about 65 to 70 on the Rockwell C scale while the parent aluminum is typically about 59 on the Rockwell A scale (a great deal softer). For example, a very good knife blade made of heat-treated steel would be just about 60 on the Rockwell C scale.

Normal piston ring materials range from cast iron and ductile or malleable iron (nodular) to stainless steel and various tool steels with some rings being chromium plated. The most popular ring material in performance applications is probably nodular or malleable iron that has a molybdenum surface treatment. The ductile and malleable iron is very tough and forgiving of heat exposure and the molybdenum surface is very hard and wears well giving good service at high piston speeds.

*Ring Flutter*—*Ring flutter* is a term that refers to the lack of stability of the ring at high piston speeds. It happens when instead of staying in contact with the cylinder wall in order to seal properly, the ring wiggles itself free of the contact. When ring flutter occurs, the ring skips and doesn't stay sealed against the cylinder wall allowing the leakage of combustion pressure to pressurize the crankcase volume. Ring flutter is also the result of detonation when the vibrations shake the ring loose from the cylinder wall. When the ring loses contact with the cylinder wall, it allows more oil into the combustion chamber and causes more detonation and the problem is multiplied. Ring flutter is often a problem at high rpm and the forces of acceleration on the piston (and rings) becomes high and the clearance (vertical) in the piston ring groove can add to the problem. Flutter is generally decreased by using correct ring clearances and less mass (thinner rings). Sometimes the ring flutter can be such a destructive force that the rings will break and damage the cylinder walls and pistons. It is a common practice in drag racing engines to use *gas ports* in the piston to keep the ring loaded, which decrease the flutter phenomenon.

In a spark ignition engine, if more oil gets into the combustion chamber, detonation can occur because the fuel becomes normal fuel mixed with oil (very low octane). Diesels are a bit more forgiving about the extra oil in the chamber, but it is still not a desirable condition for the engine.

*Blowby*—*Blowby* is the leakage caused by the piston rings. The blowby of an engine that is sealing well is generally worse at low speeds and seals better at higher rpm (under load). A poor ring seal is when the blowby value is above 2 cfm when the engine is under load. It is not uncommon for a well-sealed engine to have only 1 cfm of blowby at high power and high rpm. Ring flutter increases blowby considerably. Blowby will cause an immediate loss of power as combustion gases escape into the crankcase. The phenomenon of detonation in the combustion chamber typically will shake the rings loose and then the rings go into flutter. Sometimes a static cylinder leakage test (with a

# Basic Engine Terms, Concepts & Components

A quality leakage meter should be in any gearhead's toolbox when it is required to keep track of ring seal or setting injectors or other duties. This is a worthwhile way to keep track of an engine's health so you can tell about valve sealing or ring leakage. The testing process is static instead of dynamic, but provides a good checking method.

properly calibrated leakage meter) will help to find if the rings are leaking or the valves are leaking. A compression test can also help to identify major problems.

According to some experts on rings and ring sealing, up to 25% of brand-new rings are not up to the specifications required for minimal blowby. How much power can be gained by paying attention to these sorts of things? Generally 1% to 2%, and sometimes more can be gained. The extra efforts pay off in small increments, but the difference is often seen at the finish line or in other ways that reflect engine efficiency. Critical things such as bore concentricity, bore finish and bore distortion under load can have great effect on the potential ring seal of the cylinder.

The blowby in a running engine can be measured on a dynamometer with any number of techniques. In order to get an accurate blowby reference the engine must be loaded on either an engine (preferred) or a chassis dynamometer. Some blowby measurements have been made on the road with specialty onboard data acquisition systems, but this is rather expensive. The main issue is to rate the blowby in a dynamic fashion.

## Piston Speed

Piston speed is normally expressed in feet per minute (fpm). We will detail how to make this calculation elsewhere in the book, but for now the equation is:

$P_s = (S \times rpm) \div 6$
where $P_s$ = **Piston speed in feet per minute, S = stroke in inches, rpm = engine revs per minute**

The two examples of the largest and smallest engines described on page 3 can be compared by looking at the piston speed to equate the two engines. The larger engine has a stroke of ~98 inches and at 102 rpm the piston speed is 1,666 feet per minute. The small engine has a stroke of 0.3125" and at 26,000 rpm the piston speed is 1,354 feet per minute. So this is interesting to see that it is one way to compare the two engines. The tiny engine has a power density of about 108 Bhp (brake horsepower) per liter (1.77 Bhp/in$^3$) and the large one has a power output of only 4.27 Bhp/L (0.07 Bhp/in$^3$). These methods of evaluation are referred to as normalization and are convenient for comparisons. On the larger engine, the torque per cubic inch is an awesome 3.604 lbs-ft/in$^3$. Most performance engines are rated at much less than half that amount per cubic inch.

## Compression Ratio

The compression ratio (CR) is the ratio of the volume of the engine at bottom dead center (BDC) to the volume at top dead center (TDC) with the cylinder head and head gasket in place. This is known as the static compression ratio or mechanical compression ratio. The compression ratio is calculated by the following equation.

$CR = V_1 \div V_2$
where **CR = compression ratio, $V_1$ = volume of the engine at BDC (includes chamber, deck, piston reliefs, and volume of cylinder, and gasket), and $V_2$ = volume of the engine at TDC (includes the chamber, deck, piston reliefs and gasket)**

Normal compression ratios are typically within the range of 7:1 to 11:1. Compression ratios in excess of 10.0:1 make the engine very sensitive to fuel octane and spark timing. Naturally aspirated racing engines might have compression ratios of up to 14 or 15:1. Diesel engines typically have compression ratios above 15:1 and 20:1 to 22:1 is very common. The dynamic compression ratio is a number that takes into consideration the conditions when the intake valve closes. It is very common for the dynamic compression ratio to be many points lower than the static compression ratio.

Regardless of the size of the engine, the relationships of pistons, pins, connecting rods and other parameters such as piston speed are common ways to evaluate and equate engines. You can calculate piston speed in any engine by knowing the stroke and rpm. Photo courtesy Conley Precision Engines.

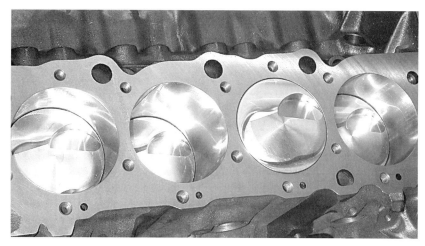

The pistons fill the bores and the bumps on the pistons raise the compression ratio (CR). These pistons are 10.2:1 units with the proper cylinder heads in place. Notice the smooth cylinder walls in this short block assembly. See the text for comments on ring seal and flutter.

The connecting rods in the photo to the left are for various applications. The photo to the right shows that connecting rods come in all sorts of shapes and sizes. One common issue of connecting rods is that they have a specific length that helps to establish where the piston is located relative to the crankshaft position.

Supercharged engines will have a higher dynamic compression ratio because of the additional pressure in the intake manifold at positive boost levels.

## HUCR

This is the abbreviation for *highest usuable compression ratio*, which is dependent upon what particular fuel is being used and the combustion chamber design. Although this is not a very common term in today's jargon, it still has application when discussing fuels and combustion. Today this number is sometimes called the *critical compression ratio*. See Chapter 2 and page 148 for details. Because of today's poor quality pump fuel (gasoline), the HUCR is from about 8:1 to less than 10:1 (depending on elevation and local barometric pressure). With the increased availability of E85, the HUCR is greater than 11:1. The formula to find an HUCR number:

$HUCR = (T_{auto} \div T_{amb})^{2.5}$
where HUCR = highest usable compression ratio, $T_{auto}$ = temperature of auto ignition of fuel in degrees F + 273, $T_{amb}$ = ambient temperature degrees F + 273

You need to consider other details such as spark advance selection and inlet temperatures as well. Those numbers are typically found after extensive testing on a dynamometer. The thing to consider here is that the compression ratio is an important part of the mix that makes up the characteristics of the engine. The engine's airflow and its dynamic compression ratio are more important, but the static compression ratio is more often an easier number to come up with because it is easily measured from the volumes involved.

## Head Gasket

The head gasket is squeezed between the cylinder head and the cylinder block when the three components are bolted together. This critical component (head gasket) is also referred to as a "fusible link" that has a tendency to blow out and leak under severe detonation. This is somewhat of a safety valve of sorts, but the gasket needs to seal properly in order to continue to help the engine produce power. Most engine designs use head gaskets of various materials that range from fiber to steel and copper. Some engines have the cylinder head and the cylinder made together without using a gasket to separate them. The no-head-gasket design is commonly used in aircraft engines, was once popular in the famous Miller, Offenhauser and Meyer-Drake engines used in early Indianapolis 500 race cars. The engine can't experience a blown head gasket if it doesn't use one. Worthwhile to mention here is that if the tune-up is correct and the spark and fuel settings are correct, the head gasket will not blow out. It might not even need O-rings in the block or heads or other special applications designed to keep a head gasket in place. If an engine is having head gasket problems, this usually means something is not right in either the spark advance or the fuel mixture, or both.

## Rod Length

*Rod length* refers to the length of the connecting rod measured from the center of the piston pin end to the center of the crankpin end, in inches. The length of the connecting rod establishes (geometrically) how much friction is imparted to the cylinder walls and has an effect on the time

# Basic Engine Terms, Concepts & Components

The deck height here is measured at just 0.005" below the surface of the block when the piston is located at TDC. When used with a 0.038" head gasket, the distance between the flat of the piston and the flat of the cylinder head is only 0.043" and that gives just a bit of room for "rock-over" at TDC when the engine is running.

(degrees of crankshaft rotation) that the piston spends at TDC and BDC. This timing is also affected by the piston pin offset if any is used. The question of whether a long or short rod is better has been an ongoing argument among engine builders for years. If you are new to internal combustion piston engines, understand that the rod length does not have an effect on either the displacement or the stroke of the engine. One half the stroke plus the rod length plus the compression height of the piston establishes the relationship of the location of the top of the piston to the top of the deck of the block.

## Deck Height

*Deck height* normally refers to the distance from the top of the piston (flat portion) to the top of the cylinder block. This dimension can be below the block (*negative deck height*), even with the block (*zero deck height*), or above the block (*positive deck height*). **Hint:** The piston must not impact the cylinder head when the engine is running. About as close as it can get without impact is the goal and the piston-to-head distance is typically something close to a range of 0.030"–0.060". This range depends on both piston skirt-to-wall clearance and whether aluminum or steel connecting rods are used. The skirt-to-wall clearance varies the amount of piston rock-over that the piston can experience. Of course the head gasket thickness is part of the measurement to be considered in the critical piston-to-head clearance issue.

## Block Deck Height

The term *block deck height* refers to the distance from the center of the main bearing bore to the top of the block. This dimension is normally given in inches to at least the third decimal place as in 8.875", 9.025" or 10.000" or whatever the dimension is measured. The block deck height minus the assembled lengths of one half the crankshaft stroke + the compression height + connecting rod length will produce the answer for the deck height. This dimension should be verified by trial assembly. Careful consideration must be given to the thickness of the assembled head gasket to complete the overall deck height. The block deck should be the same when measured from the centerline of the main bearing bores to the top of the block, verifying the deck is parallel to the main bores. Reference to the block deck can also mean the surface where the cylinder head(s) mount and the area that is separated from the head by a head gasket.

## Cylinder Block

The block places all the cylinders in location to the main bores. The basic strength and geometry of the cylinder block defines the mechanical strength and the capability of the engine. The crankcase is the volume of the engine that is covered by the oil pan and refers to the area and volume in the vicinity of the crankshaft. The crankcase volume might also be considered to be all the volume below the pistons. The most common material for cylinder blocks is cast iron although aluminum alloys are also available. The block establishes the mechanical strength foundation that the engine is built upon. Some designs are not suitable for high-power modifications, so take a close look at the block before you start pumping cash and effort into what might end up being an endless pit of frustration.

## Bearings

There are several bearing designs used in engines. Roller and ball bearings are very commonly used in motorcycle engines and some exotic cars, but the most common are the insert type for both the main and rod bearing journals. One-piece sleeve bearings are used to support the camshaft in most engines. Some overhead cam applications have the camshaft running directly on the aluminum saddles in the head. The bearing materials vary, but are most common in multi layers of tin, cadmium, lead, aluminum, and bronze with steel backings. The job of the bearings is to keep oil in place so the crankshaft or camshaft does not come in contact

# Engine Airflow

**The importance of engine bearings is one of the many basics that should be learned early on. A hydrodynamic "wedge" of oil between the bearing and the crankshaft surface is less than 0.001 inch in thickness when loaded, yet it magically helps keep the engine running even at high loads. You should also learn how to take precise measurements of all bearing clearances.**

**The curtain area is the area exposed between the valve and the seat. A very large curtain area is shown but the visual concept is helpful. Making a sketch helps.**

with the bearing surface. The oil provides a hydrodynamic wedge that has amazing capability. If the crankshaft ever comes in contact with the rotating shaft, instant bearing failure is often the result. The best thing to do here is depend on the experience and knowledge of the manufacturers and your engine builder. Simple things such as selecting the correct bearing clearances often make the difference in a solid and reliable engine and professional experience goes a long way in these applications. There are complete books written on bearing designs and applications that are worth reading if you want more detailed information.

## Port Velocity

This term references the velocity (speed) in the port or runner that is being tested. The equation for the calculation is simple to apply. This equation gives the mean or average speed in feet per second. For a reference, 60 mph = 88 ft/sec.:

$P_{vel} = (P_s \div 60) \times (B^2 \div A_p)$
where $P_{vel}$ = port velocity in feet per second, $P_s$ = piston speed in feet per minute, B = bore diameter in inches (squared), $A_p$ = area of port in square inches.

So, if an engine has a port velocity of about 300 ft/sec that equates to a little over 204 miles per hour! That is a pretty serious internal wind going down the tunnels and tubes! It is worthwhile to mention here that the boundary layer inside the port varies from about 0.030" to about 0.080" in thickness. Look for more discussion of this issue later on. Port velocity is discussed in greater detail on page 61 and in Chapter 9.

## Boundary Layer

This is the layer of reduced velocity in fluids, such as air or the airflow within a port or runner. The *boundary layer* is directly next to the surface of the port runner wall roof, and floor that the fluid is flowing past. This term can also be applied to the layer identified as the boundary layer on the outside of a vehicle or device that is going through the fluid (in this case atmospheric air). The thickness of the boundary layer varies throughout the port or runner, but is commonly about as small as 0.030" to 0.080", depending on other conditions concerning the shape and surface and obstructions. Normally, the very shiny and finely finished surfaces have tighter boundary layer conditions than do rougher surfaces, and for this and other reasons, highly polished ports have become less popular among top engine builders.

## Curtain Area

This is a term that is somewhat confusing until you understand how it applies. The *curtain area* is literally the area between the valve and the seat at any given lift position. The calculation for the curtain area is:

$A_c = \pi \times D_v \times L_v$
where $A_c$ = curtain area, $D_v$ = diameter of valve, $L_v$ = lift of valve.

Example for a 2.00" diameter valve at a lift of 0.500":

$A_c = \pi \times D_v \times L_v$
$\pi \times 2.00" \times 0.500"$ = curtain area ($A_c$) of 3.2526 square inches

Essentially this is a method to see at what lift the valve needs to be to get to the same area as the port

## Basic Engine Terms, Concepts & Components

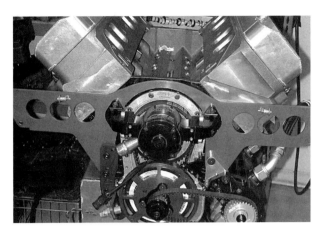

This 600 cubic-inch monster ended up making a tad over 1,200 hp on the dyno. All the components have to work together for results like this to occur. Lots of airflow preparation and planning was expended on this motor. The manifold is missing in this photo because it was removed to finish installing the roller lifters prior to dyno testing.

Even something as old as this carefully rebuilt 1929 Auburn needs strict attention paid to the details of airflow in and out of the engine. When it is finally restored this sweetheart will be worth stacks of money and will run like a Swiss watch.

This 426 Chrysler Hemi that has been completely refurbished and modified inside to provide more airflow and power with economy if the owner will keep his foot out of it. Even though Hemis don't have a very excited combustion chamber in stock form, they can be "fixed."

cross-section so that the airflow is not limited by the valve.

You can make it much more complex than the simple relationship described above by including an allowance for the valve and seat angles and the width of the valve seat, but it only adds more math without really telling you much more. But for those math geeks out there, the formula is:

$A_c = \pi \times (D_v - W_s) \times 2 \times \cos(\alpha)$

where $A_c$ = curtain area, $D_v$ = diameter of valve, $W_s$ = width of seat, $\cos(\alpha)$ = cosine of the angle of the seat (see cosine chart on page 147).

Intake port cross-sectional area *minimum* is calculated easily as:

$I_{pca} = cfm \div 146$

where $I_{pca}$ = intake port cross-sectional area in square inches, cfm = cubic feet per minute airflow at 28 inches of water (in.$H_2O$) test pressure

Engine rpm at peak power based on cross-section is not always something that is very intuitive, but this has a tendency to set the characteristics of the engine power curve. There are other calculations for the minimum cross-section in intake ports that is presented in the intake and exhaust tuning section, beginning on page 106.

### Torque/Cubic Inch (lb-ft/in³)

The relationship of engine torque to airflow is something that we will be studying in this book. In general, naturally aspirated gasoline burning internal combustion piston engines will produce torque numbers of about 1.25 lb-ft/in³ to 1.35 lb-ft/in³. This is for well running parts combinations. Higher compression ratios and very highly tuned and more efficient engines might produce as much as 1.5 lb-ft/in³ to 1.55 lb-ft/in³. In order to achieve these higher outputs of torque per cubic inch, the engine configurations would have to have complimentary cylinder heads and camshaft and header selections as well as very good mixture motion and efficient combustion processes. Normally the more efficient the engine becomes, it should require less spark advance (BTDC) to achieve complete combustion. However that is dependent on several factors such as bore size, spark plug placement, fuel used and combustion chamber shape. Hemispherical and pent-roof combustion chambers of three- and four-valve cylinder heads (relative to airflow) are the most efficient. However, true hemi-type chambers have some definite problems associated with having adequate mixture motion. The lack of mixture motion in the combustion chamber leads to less complete burning from the piston to head clearances being too large

Rich enough to suit you? Clark-Smith-Leggitt blown gas lakester runs plenty rich at start up and throughout the day. It owned the record at Bonneville at 309 mph. Shortly after this photo was taken, it recorded a smooth and safe 304 mph run. Photo courtesy Larry Ledwick.

Lean enough to suit you? If you lean it out too much, the combustion chamber will turn other things into consumables even if it is self-destructive. Aluminum can become fuel in these cases and it melts but doesn't make much horsepower. Think hard on that issue.

and they have a poor "squish to volume" ratio resulting in low squish velocities. For a definition of squish, see page 21.

### Rich and Lean

No discussion of the tuning aspects of engine performance would be complete without presenting some comments on burning the fuel the engine uses. Lifetime industrial arts educator Larry Schneider is also a Bonneville competitor and he says, "Just like a racing budget, start out rich and go from there…"

The configuration or tuning point for best power is typically realized at between RBT and LBT (*rich best torque* and *lean best torque*, respectively). Both of those points can also be established in properly applied dyno testing procedures. Never forget the axiom and warning that "it is always better to be tuned 5% rich than 1% lean."

Those simple words can save you a lot of grief and help you make a lot of usable horsepower. I can't stress enough the importance of running enough fuel to make power safely. The breadth of the span of RBT to LBT varies with each type of fuel. Gasoline-based fuels are the most narrow and least forgiving if you venture into the lean zone.

Sir Harry Ricardo (1885–1974) was an engine researcher whose book, *The High Speed Internal Combustion Engine* is a highly regarded reference. He also founded Ricardo Consultants, which is still in operation as of this writing. He was quite an icon in the field of engine design with a specialty in combustion. He found that you could actually encourage detonation to occur if you ran an engine to the point of 60% rich, but it only takes a tad past stoichiometric (chemically correct) on the lean side to start the death rattle of a blown engine.

Another quote is from former racing engine builder and engineer Bill Hancock, who said, "In my humble opinion, pistons melt to provide another data point on the surface map defined by three axes: lead, mixture and load. As we develop this map, there are those who will use it to advantage and those who will continue to contribute valuable data points. They will rationalize these contributions by claiming to define the edge."

# Chapter 2
# Internal Combustion Engine Fundamentals

Ken Weber points out that when searching for the last little bit of power or economy in an engine one must look a lot and think even more. Looking at the mold of a port helps to sort out some questions and answers. Even sketches help to think how things work.

*Do what you can, with what you have, where you are.*
—Theodore Roosevelt

In with the good air and out with the bad air is a pretty simple premise, but it is nevertheless quite accurate on how internal combustion (IC) engines work.

## Basic Internal Combustion (IC) Concept

The engine as a self-driven air pump is a fundamental concept you need to understand in order to fully appreciate how internal combustion piston engines work.

An IC engine can't ingest air on its own and that is why a starter is used to get it into initial motion so that as it fires and begins running. After that, it has an intake cycle where it takes in fresh air.

Fuel is provided and the burning of the fuel (with the ingested air) pushes on the piston and rotates the crankshaft. Some of the energy between firing cylinders is stored in the flywheel and the rotating components and supplies enough energy to expel the exhaust gases and then the process starts all over again. Simplistically described, this is somewhat the method of operation for a gas turbine, too. Fuel, air and ignition in the correct proportions and at the right time allow the engine to burn the fuel and put some of the energy that is liberated to work in powering whatever vehicle the engine is placed in. An internal combustion piston engine will not run in a vacuum.

The efficiency of the combustion process is part of the conversion of liquid fuel into usable power. That efficiency is normally reflected by the brake specific fuel consumption or BSFC number when tested on a dynamometer. The BSFC

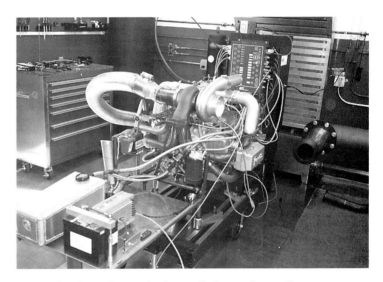

This turbocharged pancake four-cylinder engine on the dynamometer is just another self-driven air pump but it inspires the imagination and interest of gearheads everywhere. Photo courtesy Dynamic Test Systems.

number is normally expressed as fuel use in pounds per horsepower hour (lbs/hp-hr). Whenever we refer to the efficiency of the engine with how much air that it consumed, the reference would be BSAC or brake specific air consumption.

# Engine Airflow

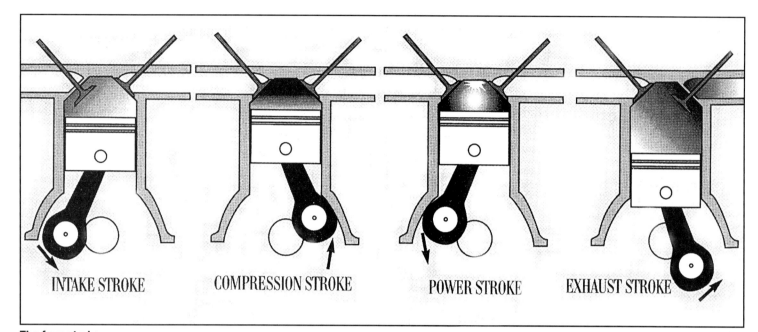

The four-stroke cycle of the internal combustion engine. Shown is a hemi-design combustion chamber, with a clockwise rotation from the rear. Courtesy Mike Kojima.

## The Four-Stroke Cycles

There are four separate strokes and definitive cycles in a four-stroke cycle engine, known as the Otto cycle. They are easily identified as: intake, compression, power (sometimes referred to as combustion), and exhaust. In street-speak the events are sometimes referred to as suck, squeeze, bang and blow—if that helps you to remember them better.

It takes two complete revolutions of the crankshaft to complete all four cycles. Conversely, a two-stroke cycle engine completes each cycle once per revolution.

Again, the evaluation for the four cycles of a four-stroke cycle spark ignition engine is in the following order. This is the classic Otto cycle for an internal combustion piston engine.

*Intake Cycle*—Fresh air from the ambient surroundings is available to push into the engine as soon as the intake valve opens. That's right, it is *pushed*, not drawn in. It is pushed in by whatever the atmospheric pressure happens to be at the time. What is surprising is that the amount of air pushed in is the same at sea level as it is at the top of Pikes Peak (14,110 feet). What is not the same is the density (or amount of oxygen) of the air, which has a drastic effect on the IC engine. Remember, the IC engine is a positive displacement air pump. The power output of the engine is directly dependent upon the density because the oxygen in the air is necessary to help burn the fuel for combustion. It is worthwhile to note that of the four valve events, the intake closing is one of the most important and helps to define the dynamic compression ratio and many of the other engine characteristics. The power

With this type of valve arrangement the process can get off to a great start with three intakes and two exhausts. There are all sorts of wild things out there if you just look around. VWs and Audis use these on turbo models. Photo courtesy Endyn.

produced during this phase is really determined by the critical relation-ship between cylinder pressure and cylinder volume.

*Compression Cycle*—As the air gets into the engine, so does some fuel, and hopefully in the correct ratio. The fuel and air are either mixed in a carburetor or the fuel might be injected into the air stream from electronic injectors or mechanical injector nozzles. If the air and fuel mixture is the correct mix, after the intake valve closes, the air and fuel will be compressed into a much smaller volume into the combustion chamber. Compressing the larger volume into the smaller volume of the combustion chamber generates heat, which preheats the mixture and helps to vaporize it, or at least make the fuel droplets smaller. The air and fuel mixture at this point is quite flammable if ignited. And somewhere up toward the top of the piston's travel in the cylinder, the ignition sends a spark, igniting the fuel and air mixture, which begins as a small flame front that continues to burn the expanding mixture as it goes to the next stage or cycle of events. If you have a problem visualizing the compression and heat relationships, rapidly pump a bicycle air pump or a hand pump for a basketball, and touch the cylinder after a few hard strokes. All the real excitement in the engine happens after the intake valve closes and the compression process begins to start the chain of events that leads to combustion. See the section on in-cylinder pressure measurement, page 22, for a more complete presentation on this cycle of engine events.

*Power Cycle*—This cycle is also referred to as the *combustion, ignition* or *expansion* cycle. An ignition source (spark) starts the internal flame that allows the burning mixture to be in full combustion mode. The pressure in the cylinder increases steadily as the gases begin to burn, so that the expanding energy pushes on the piston. As it does so, the force is transferred to the crankshaft via the connecting rod. It is that twisting force that will pass on torque to the flywheel of the engine. The difficulty in all this is the expanding burning gas pushes on the piston and the pressure in the cylinder decays from the peak pressure that occurs somewhere around 12 to 20 degrees after top dead center (ATDC). Somewhere down near the bottom of the piston travel, the exhaust valve will begin to open and the next cycle will begin. It is worthwhile to note that the combustion process is such that the fuel and air mixture burns at a controlled rate and does not "explode." If the mixture explodes, all sorts of engine damages are the result. Smoothly *burning fuel* helps to make power. *Exploding fuel* just destroys parts.

*Exhaust Cycle*—Because the cylinder is still at a fairly high level of internal pressure from the expanding and burning gases when the exhaust valve opens, the exhaust gases begin to escape fairly rapidly and the exhaust "blowdown" of the cylinder begins. That is where the exhaust note comes from, which many gearheads and enthusiasts find to be exhilarating. Think of the exhaust cycle as one that also helps to set the character of the engine's performance curve. If the exhaust valve opens too soon, it will allow wasted energy to go out the exhaust tract instead of pushing on the piston to make power. If the exhaust valve opens too late the engine has to work against itself and excessive pumping losses occur. The best point of exhaust valve opening is when the expansion of burning gases is just about done and a balance of pressure loss and correct timing is the result. Early opening exhaust valve timing does make impressive noises as the burning continues in the exhaust tubing and exhaust system, but that doesn't necessarily make power. Do not associate noise with power. Normally the pressure in the cylinder at the point of exhaust valve opening (EVO) is much higher than you would expect (70 psi to above 200 psi is not uncommon) so the exhaust blowdown and rapid decrease in cylinder pressure occurs very quickly.

As mentioned, it takes 720 degrees of rotation to complete the four cycles, and all the cylinders in the engine must go through the same process. Time after time, the cycles continue until the ignition is

Working hard on the Cummins six-cylinder diesel pays off when you put in the extra effort. This cylinder head is particularly problematic on the intake side. Diesels respond well to porting and flow bench testing. Photo courtesy Lawrence Machine.

This Bonneville streamliner will use two of Lawrence's Cummins six-cylinder engines. The liner went over 306 mph with only one engine and Roy Lewis and his crew expect much more with the new engines and airflow work.

This small twin-cylinder Honda motorcycle engine improved a great deal after a small amount of airflow work and a decent valve job. All engines will respond to improvements such as this. This cylinder head is being tested on a flow bench.

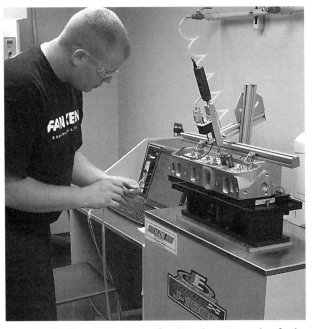

This student is assigned to map an intake port in a small-block Ford cylinder head. This well equipped flow lab would spoil anyone as they learn the details about airflow and engines. Photo courtesy Ranken Technical College.

> ### Hale's Seven-Step Theory
> Having listed the four strokes of the internal combustion piston engine in the classical sense, I would like to share with you a different perspective. In his book on engine simulation software, Patrick Hale stated that a four-stroke engine has seven separate steps that make up the traditional four strokes. They are:
> 1. Intake pumping
> 2. Intake ramming
> 3. Compression
> 4. Fuel burning and expansion
> 5. Exhaust blowdown
> 6. Exhaust pumping
> 7. Valve overlap.
>
> Hale considers these seven events as part of a chain that must always work together in order for the engine to perform well. As we get into engine airflow theory a bit deeper, it will become clear that Hale is on to something.

switched off. In the case of a diesel engine, the fuel must be turned off in order to make the engine stop rotating. That is the same way that a nitromethane fueled engine has to be shut down. Turn the fuel off and the combustion stops because the fuel supply and much of the oxygen has stopped.

*The Two-Stroke Engine*—A two-stroke cycle piston engine is much simpler, in that it has a power stroke every rotation of the crankshaft. The power density is generally expressed as power per cubic-inch of displacement. Two-stroke engines typically have a greater power density than four-stroke engines. Two-stroke engines were invented by Sir Dugald Clerk (1854–1932) in 1878. Clerk wrote a book in 1896 on internal combustion piston engines, which focused mainly on the two-stroke engines at the time.

*Diesel Engines*—With four-stroke diesel engines, spark is not created as a source of ignition. Instead, fuel is injected directly into the combustion chamber at a point late in the compression cycle, and because of the heat from compressing the air (sometimes a compression ratio of 22:1 and greater), the fuel ignites and combustion begins for the expansion cycle. The diesel or compression ignition engine was designed by Rudolf Diesel (1858–1913) in 1892 and was originally intended to burn coal dust for fuel. The coal dust scheme did not work out well and various distilled petroleum spirits were tried with much more success. Today's diesel fuel is nothing like that which was used in the early days for this engine design. Diesel fuel today is much more refined than at any time in history and as a result helps to make the diesel a very efficient engine while maintaining control of exhaust gas emissions. However that extra refining and other elements have also driven the cost at the pump higher and higher.

## How Fast Does All That Cycle Stuff Happen?

As stated before, it takes two complete revolutions of the four-stroke engine for a completion of all the four cycles to occur. During the intake cycle, the intake valve opens and closes. During the exhaust cycle, the exhaust valve opens and closes.

At 6000 rpm, each intake valve opens and closes 50 times per second! During that same one second, the piston reversed directions at TDC and BDC 25 times. And that was just at only 6000 rpm.

Here's how we calculate that speed. It is a given that 60 rpm is also equal to one revolution per second (rps). So 6000 rpm ÷ 60 rps = 100, and since it takes two revolutions for one complete cycle, the answer becomes 50. The 60 rps can also be called 60 Hz (Hertz) because it is relating to the events per second.

Now let's put this newfound knowledge to work. How fast are the valves clicking away at 8200 rpm?

$V_e$ = (rpm ÷ 60) ÷ 2 = rps
where $V_e$ = valve events, rpm = engine speed in revolutions per minute, rps = revolutions per second

$V_e$ = (8200 ÷ 60) ÷ 2 = 68.33 times per second or 68.33 Hz (Hertz is a reference to cycles per second)

## Inside the Combustion Chamber

The shape of the combustion chamber in an engine can be simple or complex. Most of the time, the overall design of the combustion chamber is a result of the footprint required for valves and spark plugs. In the case of diesels, the top of the piston is typically the combustion chamber and the cylinder head itself is actually flat, and it might also include the design of the top of the piston as well.

Another important feature of an efficient combustion chamber is the proximity of the piston to the cylinder head at TDC. If this dimension (called piston-to-head clearance) is pretty tight, the effect is that the chamber will be less prone to detonate end gases because the close proximity produces a great *squish* velocity so that the gases are forced out into the combustion chamber for proper burning. This turbulence is called *mixture motion* and is highly desirable.

***Quench & Squish***—There are a couple of terms that you should be comfortable with concerning the combustion chamber. Those terms are *quench* and *squish*. The squish of the combustion chamber is the flat area of the piston as it becomes very close to the cylinder head at TDC and is around the outside of the piston to cylinder head. This squish area pushes the mixture toward the center of the combustion chamber by what is referred to as *squish velocity* and in general adds a great deal of turbulence to an otherwise stagnant mixture, making it much easier to burn. A high ratio of squish area to volume of the combustion chamber helps to provide more complete combustion of the fuel and it helps to decrease the possibility of detonation.

The *quench zone* is the very thin cross-section, and the expression is often used in place of squish. It has been shown in testing that the quench

John Mehovitz built this twin-turbocharged Ford 4.6L (281 cid). It is all about airflow and smooth combustion. This amazing engine produces over 1600 hp on the dyno, and is a record holder, going as fast as a 6.62 ET at 218 mph. A more recent car has gone 6.49 at 229 mph. Photo courtesy Westech Performance Group.

(squish) area should place the piston a distance from the cylinder head calculated by multiplying 0.005" by the bore dimension. For a 4" bore engine, that would be 0.020". While this measurement will allow less detonation of the end gases, it is somewhat close mechanically when looser skirt to wall clearance is used. It is pretty common on high rpm 4" bore engines to seek approximately 0.025" to 0.030" between the piston flat and the cylinder head at running temperatures with steel connecting rods. Aluminum connecting rods generally require more space for material growth.

The flame speed inside the combustion chamber is variable relative to the type of fuel that is used and the compression ratio and the design of the combustion chamber too. If the fuel is one of the many in the gasoline families, the flame speed might be as high as something on the order of 500 feet per second at 6000 rpm. In a 4" bore, that would take about 0.00066 seconds to pass the flame totally across the space. A larger bore would typically take a bit longer in time so it is not uncommon for the larger bore packages to require a small amount more spark advance if all else is considered equal. If you have a super high compression ugly dome on the piston, it disrupts the smooth travel of the flame front, and a different ignition advance number would be required. We will discuss how to properly time the ignition spark point for an internal combustion piston engine.

Really the very best way to evaluate what goes on inside the engine is to use a device for in-cylinder pressure based combustion analysis while the engine is run on an engine dynamometer. This type of device requires in-cylinder pressure measurement and a shaft encoder and sophisticated computer data acquisition in order to use the data and turn it into useful information. The types of sensors and

# Engine Airflow

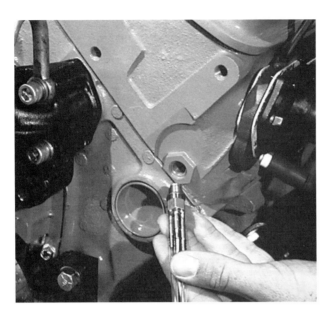

For a view inside the combustion chamber, a pressure transducer can be placed so that coupled with very complex mathematics and electronics accurately show what is going on in there. The pressure measurement also provides temperature calculations.

The exact position of the crankshaft (therefore the rod and piston) is done with a shaft encoder and accuracies of as much as 1/10 of a degree allow precise data to be calculated for what is going on inside the combustion chamber after the intake valve closes. The pressure on the top of the piston creates IMEP.

instrumentation are somewhat different for different manufacturers as are the accuracy and speed of measurements and calculations.

The use of pressure-based combustion analyses provides an internal picture of what goes on inside the combustion process and can provide reliable high-speed data. This type of equipment is used to develop most NASCAR engines and Formula 1 engines, as well as some of the top Pro Stock drag engines. These engines have been tuned to the hair splitting edge of the performance envelope. It used to be a lot easier to get things done by just providing air and fuel and spark, but today it is down to separating the gains from the losses in tiny increments. You can compare the instrumentation for the cylinder pressure–based data collection system to that used for an electrocardigram (EKG) in checking out the human heart.

## Cylinder Pressure vs. Volume

As is often the case with new topics, some terms must be defined before we get to the fun things. Of the specific processes in a four stroke-cycle internal combustion engine—intake, compression, power and exhaust—almost all the truly exciting stuff happens after the intake valve closes. In the whirling and tumbling darkness within the cylinder, as the piston starts toward top dead center after the intake valve has fully settled on its seat, pressure rises rapidly during the compression stroke as a function of many design parameters. The variables include static compression ratio, rod length, stroke, piston pin offset, and ring seal leakage (blowby). During the critical period while the piston is rising and beginning the compression of whatever fuel and air is trapped inside the cylinder, there is work being done (energy loss) on the mixture. The energy required for this compression cycle is supplied by storage in the flywheel and rotating parts that make up the *flywheel effect* of the engine.

Once the piston has passed through TDC, the cylinder supplies energy back to the flywheel (expansion or combustion cycle) if the tiny spark that occurs at the spark plug has created a kernel of flame that grows to full combustion.

***IMEP***—The cylinder pressure averaged over the cylinder volume during expansion minus that during compression is called the *indicated mean effective pressure* (IMEP). If the IMEP is positive, net energy is added to the flywheel. How quickly this energy is added determines the level of power that can be used to rip the doors off some other guy's ride.

Remember that horsepower (hp) is calculated as torque x rpm ÷ 5252. If we recognize that horsepower is simply a unit that is used to describe power, similar to how feet and inches describe length, then it becomes apparent that power is proportional to torque x rpm. The engine speed (rpm) portion of the relationship corresponds to the rate at which the torque is applied. This means that torque is proportional to IMEP. So, calculating IMEP is the same as measuring the torque of a particular cylinder before it has lost anything to internal friction or used any energy to move air in or out of the cylinder (called pumping losses).

***Other "MEPs"***—Yes, it takes some of the initial power produced to whirl all the pieces around, and it is called *friction mean effective pressure* (FMEP). Pumping losses are called *pumping mean effective pressure* (PMEP). *Brake mean effective pressure* (BMEP) is what is left when the internal friction is subtracted from the initial IMEP. When all these "techno terms" get to the flywheel, the available

# INTERNAL COMBUSTION ENGINE FUNDAMENTALS

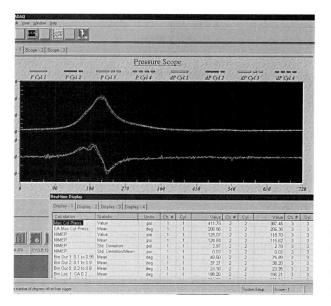

The precision of measurements and calculations that can be done by computers and data acquisition systems inside the running engine is amazing. The pressure on top of the piston tells a great deal about the condition of the engine and as these photos show, there is lots of data that can be turned into information. These type systems are not for everyone, but they are worthwhile for learning what makes engines tick.

brake torque (BT) becomes brake horsepower (Bhp) when the torque is multiplied by rpm divided by the constant, 5252.

Working through the connecting rod, which allows the force to be transmitted to the crankshaft overcoming friction, IMEP minus FMEP equals BMEP. The end result of all this force produces the twisting force at the flywheel and is ready to be put to work. Nothing is free and power has to pay the cost of the internal engine friction (FMEP) in order to get to the flywheel so that we can enjoy the effects of turning hard earned money into noise.

BMEP is what we have available to put to work propelling our vehicles. So, IMEP − FMEP = BMEP. Engine torque measured on a dynamometer can provide a number for BMEP by dividing the measured torque by the engine displacement and then multiplying by 150.8 which yields BMEP in pounds per square inch or psi. (This is for four-stroke cycle engines). More info than you wanted to know? Oh come on now, the fun is just beginning. Mathematically stated:

**BMEP = (T ÷ in$^3$) x 150.8**
where BMEP = brake mean effective pressure in pounds per square inch

T = torque in lbs-ft., in$^3$ = cubic inches displacement for the engine. The 150.8 is a constant.

Now that we've been MEP'd to the point of hurling, let's see how we can actually accomplish the most noble of goals in making maximum BMEP. Uh, that is IMEP, uh, minus FMEP and PMEP?

## Maximizing BMEP

OK, so we want to maximize BMEP, but first we need to evaluate it. If we really want to know what is going on inside the combustion process, we need to check two important relationships of the engine. We must measure cylinder pressure at known cylinder volumes and the position of the crankshaft with an electronic encoder. The technology is in place to measure these elements so that we may probe what exactly occurs when the intake valve momentarily shuts off (this happens 50 times per second at 6000 rpm and 66 times per second at 8000 rpm!).

*The ICECA*—One company makes a device called an internal combustion engine cycle analyzer (ICECA). This equipment creates a picture window for viewing the combustion process as it really happens. The ICECA equipment rapidly measures input data from sensors mounted on the engine while it's on an engine dynamometer.

The cylinder pressure sensor information is coordinated with the crankshaft position (where a shaft encoder is mounted) so that "cycle resolved" measurements are collected. The measurements are made at a rate up to 500,000 times per second for the whole system, and as many as four channels

23

# Engine Airflow

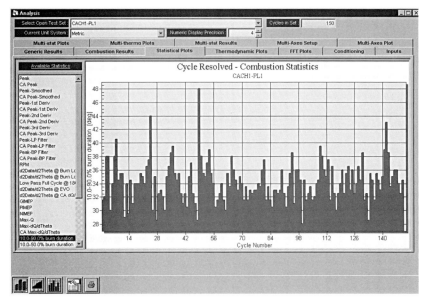

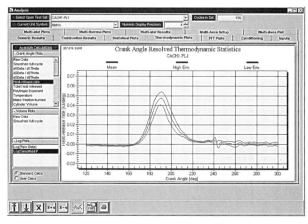

You must study the trends in order to learn how engines really function.

You would think that if a firing cylinder was measured over many cycles that the pressure would be the same. Not so. This photo shows over 140 cycles and the burn rates varied quite a bit. The text explains more details.

may be used simultaneously. That relates to a resolution accuracy of every 0.2 degrees of crank rotation up to 10,000 rpm. Special software enhancements allow the resolution to be at 0.1 degrees over the same range.

Why is so much data necessary and when is it needed? As an example, the positioning of the intake valve closing point is critical within plus-or-minus 1 degree for optimum performance. The same can be said for ignition timing and mixture strength. The strange thing is that all cylinders in a multi-cylinder engine may need different timing to produce the best results. Simply measuring with a dynamometer is not enough in the super competitive world of professional racing or professional engine builder who wants to stay on top. Imagine attempting to properly tune four engines powering one dyno and trying to find the weakest one. Well, that is only half as difficult as trying to tune eight cylinders on the same dyno.

A dynamometer measures the result of adjusting all the engine cylinders as a group. That is called *batch processing*. Measuring the power output with a dyno is an averaging compromise for the engine tuner in selecting the best valve timing, jetting, or ignition advance (or retard) in order to produce the best power without damaging the engine. The ICECA system provides the output of power of each cylinder with the analysis of its IMEP.

Similar to an electronic degree wheel, the ICECA shaft encoder continuously feeds the monitoring computer the precise crankshaft position. The typical ICECA system collects all the engine data, performs several complex calculations, and then either automatically generates color graphs for evaluation on the computer screen or printer. These graphs depict minute but critical details such as displaying the temperature of the intake valve closing event or the burn rate of the mixture (mass burned fraction). In all, there are more than 20 separate graphics displays, beginning with the basic cylinder pressure vs. volume.

Many graphs and printouts are available after running a test with the monitoring equipment: pressure crank angle, pressure-volume, log pressure vs. log volume, rate of pressure rise, heat release, mass burned fraction, polytropic coefficient, temperature vs. crank angle, temperature vs. entropy, and an overlay comparison screen.

In today's engine designs or in the refinement of an old design, more and more accuracy from within the working cylinder is paramount. This is not really a new concept, but what is new is that the information is available now and is in the hands of regular engine builders and designers instead of just in the highly secretive laboratories at the OEM level.

The selection of ring packages, compression ratios, and combustion chamber design and piston dome shapes is enhanced when viewed under actual operating conditions through the magic window inside the cylinder.

The shaft encoder is mechanically referenced to the TDC location in the initial setup of the computer. Engine displacement, rod length, piston pin offset, compression ratio, and stroke are all entered during setup so that the computer can perform its multiple and complex calculations with the collected data during the operational test.

***Testing and Data Gathering***—The first test is

# INTERNAL COMBUSTION ENGINE FUNDAMENTALS

Tough little transducers can even be used in the exhaust to select the proper header size and length. Transducers come in a variety of sizes and configurations for installation directly into a cylinder head or block. Specially modified spark plugs (least accurate because of lengthy miniature plumbing) are also available. This shows about $8,000 worth of equipment and a full system is about $35,000 to more than $50,000.

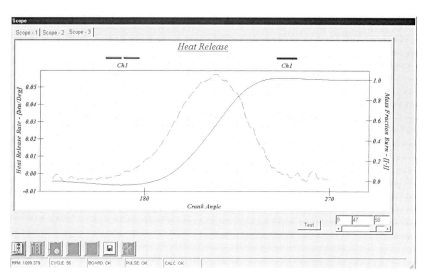

What you are looking at here is a graph of what is called heat release. The heat release is plotted in Btus per degree of crankshaft rotation. The heat release varies with many things such as fuel, mixture strength, compression ratio and other variables and it happens very fast.

conducted by "hot motoring" one of the cylinders (non-firing with spark plug cable disconnected) with the rest of the cylinders firing to provide the energy for the motoring of the one cylinder. This test verifies "pressure TDC," which may differ slightly from mechanical TDC. After this verification test, the spark plug is reunited with its wire and the rest of the series of tests are ready to be run. An interesting note on the hot motoring is that statistical data collection of perhaps 100 or more cycles *will not* produce exactly the same pressure each cycle, even though it is just pumping *unfired* air/fuel through the cylinder! The engine is an imperfect mechanical device and the goal is to improve the overall repeatability from cycle to cycle. In a firing cylinder, there are misfires (failure for complete combustion to occur) going on all the time. By the time you hear them pop, crackle and stumble, they are really misfiring badly!

The combustion process is not something you can successfully evaluate intuitively. Application of equipment such as the ICECA provides an insight into the chaos within each cylinder of the engine during combustion. For example, did you know that temperatures of over 3,000°F exist within the combustion chamber? What is the pressure in the cylinder at the point of ignition?

The proper programming of any electronic fuel injection can be correlated with the use of the ICECA equipment in order to verify the mapping process of both fuel and spark curves. LBT and RBT (Lean Best Torque and Rich Best Torque) tests can be performed so that the EFI system can be programmed to take advantage of targeting the max IMEP per cylinder (most of the more tunable units can inject fuel and adjust the spark advance on each individual cylinder). MBT (Minimum spark advance for Best Torque) tests can be run and the best timing can be accomplished with modified distributors or triggering systems. This type of test is also called a "spark hook" because of the characteristic shape of the curve when shown on a graph.

Very often, optimum tuning points can't be found because combustion knock threatens to turn the engine into a grenade. It is improbable that each cylinder will begin to knock simultaneously. It is also true that once knock is audible it is probably lethal (we're speaking of racing engines). Ideally, most engines will produce the best IMEP with the LPP (Location of Peak Pressure) at 12 to as late as 20 degrees ATDC (After Top Dead Center). The LPP is most commonly between 12 to 15 degrees ATDC on a well-tuned engine. Contrary to popular belief, more peak cylinder pressure is not necessarily better. The old racer's comment of "if a little bit is good, more is better, and too much is just right…" is wrong! Big IMEP is what we're after and that is not just higher cylinder pressure, but where things occur. That would give us a higher average BMEP. It is that area under the curve thing again. Stop thinking of peak numbers and start thinking about average numbers and increased area under the curve.

# Engine Airflow

Yep, "bottle babies" come in all ages. The popularity of nitrous oxide ($N_2O$) has been on the increase in drag racing for years. $N_2O$ is an easy way to add power but it has its drawbacks, too. Learning more about fuel and air will help you and your racing program.

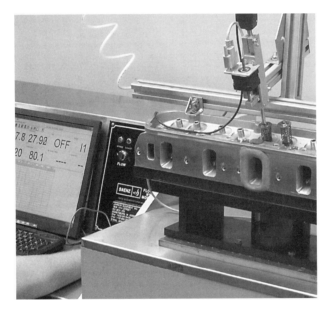

The oxygen content of the air that enters here greatly influences the power the engine can make. Learning more about the atmosphere and its effects on engines can give you specific advantages over competitors or on a project. Photo courtesy Ranken Technical College.

## Atmospheric Air Chemistry

Atmospheric air is comprised of several different gases, but the engine is primarily concerned with the amount of oxygen ($O_2$) available for fuel combustion.

The atmosphere is comprised of roughly 78% nitrogen (N), 21% oxygen ($O_2$), 1% argon (Ar) and several other minor elements. If the engine ingests less oxygen ($O_2$) then it will make less power. The $N_2$ is an inert gas and passes through the engine, absorbing energy that would otherwise be used to make more horsepower, but there's nothing we can do about that. If the air is also full of contamination from oil, unburned fuel and exhaust from other vehicles, or rubber dust, or miscellaneous bugs, the engine will also make less power. Controlling the conditions of the air, including the temperature, is one of the details that lead to free horsepower.

Conversely, if the atmospheric air has an element such as nitrous oxide ($N_2O$), and is enriched with extra oxygen (or anything else that would enhance the amount of available $O_2$), then the engine will make more power (if it has enough excess fuel to burn). On the other hand, if air had lots of carbon monoxide (CO) and unburned hydrocarbons (HC), the engine would lose power, since these contaminants would displace the oxygen.

***Density***—*Air density* and *relative air density* both describe the actual weight of the air. The fact that one cubic foot of air takes up a volume of 1728 cubic inches is something that you should commit to memory. Multiply the 1728 by 2 and you get 3456, another number you will recognize as you study airflow. The equation for relative air density is:

$RAD = ((P_b - V_p) \div 29.92) \times (520 \div T_{air})$
where RAD = relative air density, $P_b$ = local barometric pressure in in.Hg, $V_p$ = vapor pressure in in.Hg, $T_{air}$ = °F + 460

So with a barometric pressure of 29.92 in.Hg, at 60°F, with dry air that means vapor pressure of 0:

$((29.92\ in.Hg - 0) \div 29.92) \times 520 \div 520$
$1 \times 1 = 1.00$ RAD or 100%

So if you want to refer it in percentages, multiply the answer by 100.

Air Density % = $100 \times (P_b - V_p \div 29.92) \times (520 \div T_{air})$
where $P_b$ = local barometric pressure in in.Hg, $V_p$ = vapor pressure in in.Hg, $T_{air}$ = °F + 460

The weight of sea level air is considered standard at 0.0763 lbs/ft³ and of that weight, 21% of the volume amount would be atmospheric oxygen. However, what is the amount of oxygen in weight? It is pretty straightforward stuff to evaluate so we will take a quick look. From the previous description of atmospheric air, nitrogen takes up 78% of the cubic foot with 1% being various gases such as argon, and other gases. So, nitrogen (N) has an atomic weight of 14.0067 and oxygen (O) has an atomic weight of 15.9994. In the calculation, nitrogen has two atoms and oxygen has two atoms or $N_2$ and $O_2$ respectively.

Doing the calculation is 2 x 14.0067 for the nitrogen and 2 x 15.9994 for the oxygen, so the molecule of nitrogen is 28.0134 and the molecule of oxygen is 31.9988.

# INTERNAL COMBUSTION ENGINE FUNDAMENTALS

There is always a gathering of alky burners (the good kind) at the drags. The E85 interest in racing has been high because it is a good fuel that is relatively cheap for a racing fuel that is high in octane and with good burning characteristics.

This is a Cummins BT4 diesel head and as you can see it is flat where the valves are located. The combustion chamber is really in the piston. The thing you see that looks like a spark plug is a fuel injector nozzle tip.

Engines using methanol have been part of the racing scene for more than a 100 years. Since drag racing started in the US in the 1940s the "fuelers" have always used methanol as a base and they discovered nitromethane. Then things got interesting.

Then using a chemical equation relationship called the ideal gas law:

$$P \times V = n \times R \times T$$

where P = absolute pressure (lbs/ft$^2$), V = airflow (cfm), n = airflow (lbs/min), R = 53.3, which is the universal gas constant value for air, T = temperature in degrees Rankine (°F + 460)

Plugging our numbers into this equation will reduce the amount of our cubic foot of atmospheric air of 0.0763 lbs. (1.2208 oz.) to constituents of $N_2$ and $O_2$ of 0.95 oz. and 0.256 oz. respectively.

In other words, only about 1/4 oz. of oxygen is available for combustion in each cubic foot of atmospheric air at sea level. It is less at higher elevations. Believe it or not, any amount of atmospheric water that is present displaces usable gases (namely our friend $O_2$) and as a result, moist air is less dense than is dry air. So don't fall for that old deal about how "heavy" the air is when it has high humidity. That stuff is less dense than dry air and as a result will help to make less horsepower, but it does cling to you and make it seem heavy because of the water in the air that makes your clothing feel soggy.

## Chemistry of the Fuel

An engine uses fuel to burn in order to produce power and the chemical makeup of the fuel is an important consideration. All fuels have different burn characteristics that are directly related to the chemical composition of the fuel.

*Btu Rating*—The chemical makeup of the fuel is an important consideration and most fuel has a rating of Btu (British Thermal Units) per pound or per gallon. This rating is called *heating value* and can be referred as a HHV (high heating value) or LHV (lower heating value). The definition of a Btu is the following:

**42.44 Btu/min = 1 hp**
**2546 Btu/hr = 1 hp**

A more complete definition of the energy value of a Btu is that energy which is required to raise the temperature of one pound of water ($H_2O$) one degree Fahrenheit (1°F) at a pressure of 1 atmosphere (29.92126 in.Hg).

All fuels have different amounts of heating value (energy) per pound or per gallon. Typical pump gasoline varies in content and has a heating value of approximately 19,000 to 20,000 Btus per pound of fuel. It would be ideal if we could expect it to convert to power without losses, but that simply doesn't happen at that conversion rate of efficiency.

Methanol has a heating value of approximately 9,500 Btus per pound of fuel.

# Engine Airflow

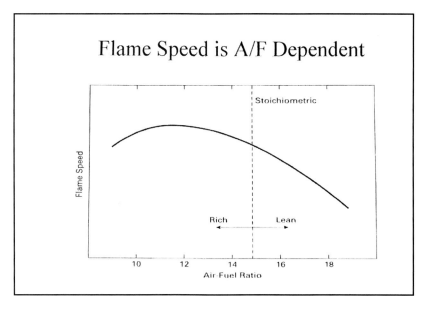

**Flame speed in the combustion chamber is an important issue on anything that revs past a few thousand rpm and the speed varied with A/F ratio changes.**

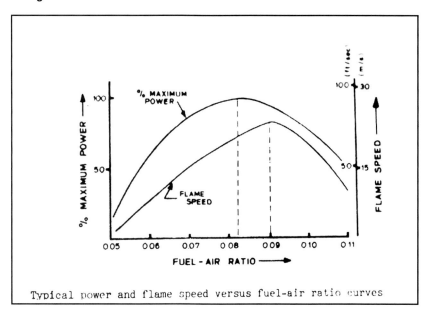

Typical power and flame speed versus fuel-air ratio curves

**Flame speed and power are plotted against the fuel to air ratio in this graph. Learning about these relationships will give you an advantage when it comes to building a high-performance engine.**

Ethanol has a heating value of 11,500 Btus per pound of fuel.

E85 (a blend of gasoline and ethanol) has an average heating value of approximately 13,475 Btus per pound of fuel. But since the ratio of gas to ethanol is so inconsistent, the actual heating value varies quite a bit.

Diesel fuel from the pump has an average heating value of approximately 19,000 Btus per pound of fuel.

*Specific Gravity*—The specific gravity (SG) of the fuel does not reflect its energy value. The SG simply refers the reference to the SG of water, which is 1.00 at 39.2°F. Another tidbit about water is that the thermal coefficient of expansion is 0.00021 per 33.8°F at 68°F, or more simply water will expand its volume about 3% from room temperature (70°F) to 175°F. So, an easy way to relate the SG number to weight per gallon is multiply the SG of the fuel times 8.34 lbs. (the weight of water per gallon at normal temperatures). A fuel with a SG of 0.850 would weigh in at about:

**0.85 x 8.34 = 7.089 lbs/gal**

Notice that that does not tell you anything at all relative to its energy content.

*Methanol*—Methanol was once made from wood, and was therefore referred to as wood alcohol. It is one of the simplest of the alcohols and today is usually produced synthetically from natural gas and a separation process. Methanol has many uses, but about 40% of our U.S. production is for making formaldehyde with a process that dates back to the ancient Egyptians. Methanol is a classic polar solvent and the products of burning methanol in air results in carbon dioxide and water. The chemical compound methanol has an octane rating of about 113. Methanol is a forgiving fuel but has several problems associated with its use and many will be addressed as we continue the investigation of fuels.

*E85*—The big buzz as this was written was the growth in manufacturing and availability of E85, which is a blend of gasoline (15%) and ethanol (85%). Ethanol is produced by converting grain corn into alcohol, snapping up about 30% of U.S. corn production at the time of this writing. The processing of corn into ethanol is a pretty simple process, similar to the one used to make "moonshine" during Prohibition. But ethanol is denatured with other chemicals so it isn't safe for human consumption.

It takes about 400 lbs. of grain corn to make 25 gallons (~164.5 lbs.) of ethanol. There are other methods of making ethanol from plants but corn conversion is the most common. It is worthwhile to also consider that it takes about 140 gallons of water to cook off one gallon of the ethanol. Technically speaking, ethanol production as an alternative automotive fuel source is not very efficient and is highly controversial. There's a lot of questions as to whether it is economically or environmentally feasible.

The octane rating of E85 varies but is generally

## Racing Gas Btu Content
*Courtesy Rockett Brand Racing Fuel*

**Summary:** High Btu content in a gasoline does not necessarily mean the engine will make the most horsepower on that particular gasoline.

Many racers and car enthusiasts ask about the Btu content of racing gasolines. They always want the gasoline with the highest Btu content per gallon. Unfortunately, this is a situation where a little knowledge is dangerous.

First, we need to define the value of a Btu. Simply put, it is the amount of heat (energy) required to raise the temperature of one pound of water one degree Fahrenheit at or near the point of maximum density. A gallon of gasoline will usually contain from 115,000 to 125,000 Btus. Most enthusiasts want the gasoline with the most Btus, and that can be misleading. The Btu content is of little value if some of the gasoline is still burning when the exhaust valve opens and all of that energy escapes out the exhaust as heat and unburned hydrocarbons. Most engines that exceed 7000 rpm can benefit from a 115,000 Btu per gallon gasoline better than a heavier gasoline that may contain 125,000 Btus per gallon that didn't completely burn in the combustion chamber. Consider this: One gasoline has 115,000 Btus and is 95% burned before the exhaust valve opens; the other contains 125,000 Btus but is only 85% burned before the exhaust valve opens. Simple math tells us that the first gasoline gave up 109,250 Btus. The other gave up 106,250 Btus. Which would you prefer?

Does this actually happen in real racing conditions? The answer is yes. Although some heat energy does go out the exhaust, some goes to the cooling system, some goes to pumping losses, etc. some of it goes into making horsepower at the rear wheels. The bottom line is that the greater the percentage of the gasoline that gets burned in the combustion chamber, the better off you are since those Btus contribute to more horsepower.

A slow-burning fuel that is still burning when the exhaust valve opens will put a flame out the pipe that can scare the dickens out of the guy next to you. In circle track or road racing, this may gain you a position by making the other guy stay at "flame length," but you could get better results by using a gasoline that burns faster, providing a higher level of thermal efficiency and therefore, more horsepower. It may look spectacular with three feet of flame coming out of your race car pipes that singes the paint on the car next to you, but those flames are energy being released in the exhaust rather than in your combustion chamber. The same thing can happen to a good gasoline if the spark is retarded. Retarded spark gives high exhaust temperatures also and contributes to overheating the engine. Think about it.

***Reformulated Gasoline***—Some people complain about reformulated gasolines (RFG), but these products have been developed to improve combustion efficiency and catalytic converter efficiency. As amazing as it may seem, some of the same racing gasoline improvements that improve horsepower in a racing engine can also lead to reduced exhaust emissions in a street engine. In both cases we want to improve combustion efficiency; in one with more horsepower, and in the other with reduced exhaust emissions.

In testing conducted using a 740 horsepower 358 cubic inch racing engine with 14:1 compression ratio, the engine made 5 more horsepower on a 100 octane unleaded RFG than on 110 octane leaded racing gasoline. How so? It is because the RFG contains an oxygenate that allows for more complete combustion and the fact that the gasolines have similar vaporization characteristics. Incidentally, Rockett Brand TM 100 Octane RFG is legal in all states including California. They have blending charts available to help you blend to a specific octane number.

---

accepted to be between 100 to 105. There are all sorts of problems associated with E85 and they will be listed as we continue our discussion on fuels.

***Hygroscopic Tendency***—Methanol and ethanol (E85) are *hygroscopic* fuels, which means they tend to absorb water from the surrounding atmosphere, altering their original chemical characteristics—water doesn't burn. When a fuel tank is vented to release pressure buildup, it also draws in outside air. That's why it isn't a good idea to let methanol or ethanol sit in the fuel tank of a race car. Make sure you keep up on the maintenance of your fuel system when running these fuels, or you'll be paying for it in parts. For more on the chemistry of these fuels, turn to page 123.

***Gasoline***—Gasoline is made from chemical refining of petroleum (oil) raw products. Gasoline is primarily made of hydrogen and carbon, so there are literally hundreds of various compounds that fall into the gasoline category. The octane rating of gasoline ranges from about 83 to 95 for pump gasoline. Racing gas has a much higher octane rating so it is substantially better from a chemical standpoint. But it is also much more expensive.

***Gasoline Chemistry***—As an example, let's examine the chemical composition for gasoline. What we would call gasoline is chemically referred to as an aliphatic hydrocarbon. The components

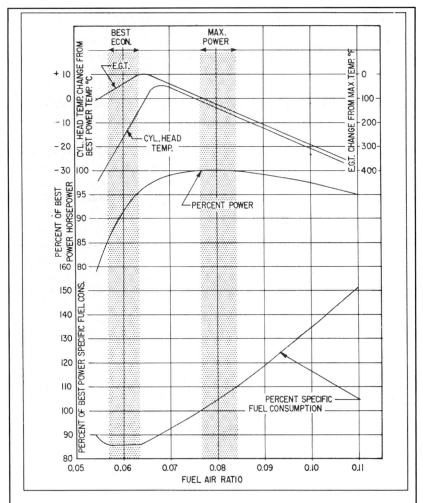

*The Lycoming representation of Fig. A-1 has idealized curves for EGT and CHT, with linear ramping to and from peak. In the real world, EGT and CHT behave more like Fig. A-1. Lycoming's graph nonetheless summarizes the qualitative relationships between EGT, CHT, power, and specific fuel consumption accurately. Notice the relationship of peak CHT to peak EGT and peak power.*

**This graphic shows the relationship of fuel-air and exhaust gas temperatures (EGT) and other variables. This is standard pilot training stuff. The reciprocal of fuel/air is air/fuel ratio. Shows clearly *not* to rely on EGTs for fuel mixture indication.**

that make up gasoline are quite varied. There are even different blends for summer and winter use.

First of all, there are hundreds of various mixtures of components that would be in the general family of "gasoline." Typical components in gasoline are methane, ethane, propane and butane. The end result is that most gasoline compounds are made up of hydrogen and carbon atoms, thus the name hydrocarbons. They are typically a chain-type structure, so they are called chain hydrocarbons. The long chain hydrocarbons ranging from $C_7H_{16}$ to $C_{11}H_{24}$ are the gasolines that end up our fuel tanks. Racing gasoline has various amounts of toluene, methyl benzene ($C_7H_8$) or xylene, which is sometimes called di-methyl benzene ($C_8H_{10}$). It also may have trace amounts of tetraethyl lead ($C_8H_{20}Pb$) and other chemicals that alter both the octane and the burning characteristics. There are even racing gasolines that are supplied without lead, and those types use other chemicals to accomplish the same thing within the design and components of the fuel. It is very common at the time of this writing that aftermarket performance suppliers of racing gasoline use MTBE (methyl tertiary butyl ether, $C_5H_{12}O$) as an octane improver and oxygenate. A gasoline compound that is typical of most high-end gasoline is iso-octane or 2,2,4 trimethylpentane ($C_8H_{18}$). The chemical balance of the combustion of tetraethyl lead looks like:

$$(CH_3CH_2)4P_b + 13\ O_2 \rightarrow 8CO_2 + 10H_2O + P_b$$

Lead was banned from pump gas in 1986. Some racing gasolines still contain some tetraethyl lead, but only in small amounts. Some manufacturers offer 100 octane racing gasoline at a condition of containing no lead at all, but use other chemical agents as anti-detonation additives.

The notion that octane is the most important consideration for gasoline is incorrect. The bottom line is an internal combustion engine needs fuel to produce power. As discussed earlier in the chapter, a more accurate method to determine the power of the fuel is to consider its HHV (high heat value), which is expressed as Btu/lb. This value is a pretty good reference of how much heat can be liberated and used per pound of fuel. The octane rating method has varied over the years and is just a way to state the fuel's resistance to detonation in engine operations. For more on racing gas Btu content, see the sidebar on page 29.

***Fuel Additives***—There are all kinds of fuel additives on the market Some might even improve the chemistry of the fuel. However, many of the additives are toxic and require special care when handling. Check the MSDS (material safety data sheet) document before handling them. Some of the chemicals are known to cause cancer and others can make you blind. So in general, be very careful what you plan to mix into your fuel tank. This is not just a lighthearted warning, as many of the additives are simply not healthy to handle.

A very common additive to gasoline is nitropropane ($C_3H_7NO_2$) in small amounts such as 5% or 6% by volume, because the nitropropane is miscible with gasoline.

It is possible to add up to 10% with positive results. However, it is necessary to plan on making use of a very rich mixture and cold spark plugs or suffer the consequences. Because the fuel carries some of its own oxygen, it needs to be run very rich in order to take advantage of the capacity to make power.

There are all sorts of additives and many of them will burn and most have problems in both handling and getting the mixture right for the engine. It is probably easier to consider either E85 or methanol or other fuels that are not additives and might also cause problems with not being legal in either racing or on the streets. When pump gasoline quality is good, it is hard to beat for simplicity in most internal combustion piston engines. Of course it has its limitations and problems too, but it is an easy choice if nothing else.

## Chemistry of the Combustion Process

Now let's put all the discussion on the chemistry of air and fuel together to see what happens during combustion. Simply speaking, the air and fuel are supposed to be all stirred up in a burnable ratio and the mixture is sometimes wishfully called a homogeneous mixture. When the ignition point is reached and a spark is produced across the gap of the spark plug (for spark ignition engines), the initial small kernel of combustion starts and the flame front begins to spread across the combustion chamber. There is a time delay from the point that a spark is created across the spark plug gap and when combustion actually begins. It depends on several things, but in general this delay is from 6 degrees to 8 degrees of crankshaft rotation. Even with spark and fuel and air in the combustion chamber in the correct amounts and timing, combustion is not guaranteed. All engines misfire occasionally. The combustion is either incomplete or not initiated at all. That might occur as much as 3% to 4% (or more) of the time. No, you can't hear or sense that level of misfires, particularly if they are random. And yes, there are partial combustion events that take place as well—it is more like a fizzle or a sputtering flame than a complete burning.

Essentially the capture of the heat from combustion is what can be converted to power by pushing down on the piston during the expansion phase or cycle. Heat is power and the more efficiently that it is captured then the more effectively the power can be generated on the minimum amount of fuel burned. Inside the combustion chamber, temperatures can reach upwards of over 3,700°F in the central core of the burning mixture. The heat is lost rapidly as it expands and the combustion flame gets down to only a few hundred degrees as it nears the thermal transfer boundary layers on the cylinder walls and the walls of the combustion chamber. The same thing applies for the piston side of the combustion process. Otherwise it would melt the metals that are in contact with the combustion event.

The photo shows a burn that is underway. Any slight change of local conditions can alter how well it burns. Think about combustion as a burn as it is not an explosion in the chamber. Uncontrolled combustion is an explosion of sorts.

***Chemical Balance of Air & Gasoline Combustion***—An in-depth study of the combustion process and the heat transfer characteristics would involve the application of the science of thermodynamics, which is somewhat beyond the scope of this book. Be we can take a look at some of the chemical issues of combustion.

$C_8H_{18}$ is *iso-octane* or 2,2,4 trimethylpentane and we will use that particular compound to represent a typical "gasoline." There are about 19,000 Btus/lb. of available energy in the fuel, but IC engines are inherently inefficient (thermally speaking), only about 25% to 30% will make it to the level of usable power. That is just about as thermally efficient as a light bulb. The rest of the energy from the fuel goes into the exhaust, oil system, cooling system and radiated as convection heat loss. The chemical balance for iso-octane is:

$$C_8H_{18} + (1.5 \times 12.5) \times (O_2 + 3.76N_2) \rightarrow 8CO_2 + 9H_2O + 6.25O_2 + 70.5N_2$$

Although the equation is chemically balanced for having available at least 150% of the theoretical air, it is a lot more confusing than simply letting the combustion chamber do all that sight unseen.

The chemical balance for a stoichiometric or chemically correct (ideal combustion) mixture of iso-octane and air is:

$$C_8H_{18} + 12.5(O_2 + 3.76N_2) \rightarrow 8CO_2 + 9H_2O + 47N_2 + \text{heat output}$$

# Flame Speed, Octane Number & Horsepower Relationship
*Courtesy Rockett Brand Racing Fuel*

There is a lot of misunderstanding about the relationship between flame speed, octane number and power.

There are some connections between these items, but not as many as some people think. We will address each one independently, then try to tie them together as best as we can. For the sake of some simplicity, the following discussion will be limited to gasoline unless otherwise indicated.

***Flame (Burn) Speed***—The speed at which the air/fuel mixture is consumed in the combustion chamber is critical in a racing engine. At 6000 rpm, each spark plug fires 50 times per second. That's a lot of combustion happening in a very short time in the same combustion chamber. That is why racing gasoline needs to be capable of burning fast. In your daily driver that may not see the top side of 3000 rpm, flame speed is not as critical. In a racing engine, everything is happening much faster, and in a bigger way because the throttle is wide open. The gasoline must burn as completely as possible to make the most possible horsepower. If the gasoline does not get burned in the time allowed, there will be unburned hydrocarbons coming out the exhaust pipe. Flame speed is determined by the hydrocarbon components in the gasoline. It is critical to making maximum power, but not related to octane quality.

***Octane Number***—The octane number of a gasoline has little to do with how fast it burns or how much power the engine will make. Octane number is the resistance to detonation. If the octane number is high enough to prevent detonation, there is no need to use a higher octane gasoline since the engine will not make any additional power. Octane number is not related to flame (burn) speed either. Variations in octane quality are independent of flame speed. There are some high-octane gasolines in the marketplace with fast flame speeds and some with slow flame speeds. It depends on how they are put together. At Rockett Brand, we like fast flame speeds because we know that a properly tuned engine will make more power on this type of gasoline than one that has a slower flame speed.

***Power***—The ultimate goal in the racing gasoline business is to convert chemical energy from the gasoline hydrocarbons into mechanical energy or horsepower. The most efficient way to convert the gasoline into horsepower is to have the correct air/fuel ratio and the correct spark timing. A mixture that is too rich or too lean will not make maximum horsepower. The same is true of spark timing: too much or too little will compromise engine output.

As indicated above, flame speed and octane number both impact the amount of power that an engine will develop, but they are independent of each other. To get maximum power from an engine, one must have a gasoline with adequate flame speed (faster is always better) and adequate octane quality to support the combustion process. Tied in with the optimized air/fuel ratio and the spark timing, we have a winner.

---

Note that there are many similarities to the previous equation. Thank goodness all this type of thing goes on without us becoming too much involved in the process.

***Chemical Balance of Air & E85 Combustion***—The chemical balance of the combustion of ethyl alcohol (ethanol) is:

$$C_2H_5OH + 3O_2 \rightarrow 2CO_2 + 3H_2O + \text{heat output}$$

The atmospheric nitrogen was left out on purpose as it is a pass-through anyway. The chemical balance for E85 using ethanol and iso-octane would look like:

$$C_2H_5OH + 0.15\, C_8H_{18} + O_2 + 3.76 N_2 \rightarrow CO_2 + H_2O + \text{heat output}$$

The RBT point of E85 will be approximately at an A/F ratio of 6.98:1 while the LBT point of E85 fuel will be approximately at an A/F ratio of 8.47:1. The E85 fuel should have the capacity to produce about 3% to 5% more than gasoline although it will use more fuel to do so. The energy in E85 averages to be approximately 84,000 to 89,000 Btu/gal. depending on the energy of the gasoline in the mix.

***Chemical Balance of Methanol/Nitromethane & Air Combustion***—Methanol ($CH_3OH$) is easily tunable for a wide range of engine packages. The span of LBT (lean best torque) to RBT (rich best torque) is fairly broad, which makes it a great choice for a racing fuel. But using methanol can also be problematic, mainly with chemical corrosion and compatibility with fuel system components. O-rings, seals and gaskets must be methanol compatible or they will be destroyed. For example, Viton-tipped needles (used in some needles and seats) are not compatible with methanol. Always check with the component manufacturers for chemical compatibility of their parts with various fuels and additives.

Methanol ($CH_3OH$) has been used in racing for

decades, going back to the early 1930s as standard fuel in midget racers, the Indianapolis 500, and at the dry lakes high-speed events. During WWII, methanol production was used for motor fuel by many countries that had more agricultural production than access to petroleum because methanol was made from wood stocks. Because that methanol carries some of its own oxygen to support combustion, it is a fuel that will make more power if tuned correctly. It is not uncommon to produce about 10% to 11% more torque (at peak torque rpm) and about 5% to 6% more horsepower at peak rpm when using methanol for fuel.

The chemically balanced equation for methanol when used as a fuel in a piston engine is:

$$2(CH_3OH) + 3O_2 \rightarrow 4H_2O + 2CO_2 + \text{heat output}$$

Nitromethane ($CH_3NO_2$) is sometimes referred to as the king of all motor fuels because of its capacity to make larger amounts of power. Nitromethane has almost enough oxygen content to be categorized as a complete monopropellant (almost needing no atmospheric oxygen for complete combustion). The chemical balance equation of nitromethane as a monopropellant (using no atmospheric oxygen) is:

$$4CH_3NO_2 \rightarrow 4CO + 4H_2O + 2H_2 + 2N_2 + \text{heat output}$$

The chemically balanced equation of when nitromethane is used as a fuel in an internal combustion (reciprocating) engine is:

$$4CH_3NO_2 + 3O_2 \rightarrow 4CO_2 + 6H_2O + 2N_2 + \text{heat output}$$

The awesome power potential of nitromethane is derived from the fact that it can burn about 8.7 times as much nitro as gasoline in one power stroke. It takes about 14.6 or 14.7 lbs. of air to burn 1 lb. of gasoline and only 1.7 lbs. of air to burn 1 lb. of nitromethane, although for maximum power potential each of the fuels needs to run richer than that (using less weight of air).

If you ever get the chance to run a bit of nitro, be very cautious and conservative, always going slightly rich with your estimates. Nitromethane has a heating value of approximately 5,100 Btu/lb.

## Airflow and Fuel Flow (Stoichiometric Ratio)

Because the engine is a self-driven air pump, the more air it pumps, the better its capacity to make

### Lambda ($\lambda$) vs. Air/Fuel Ratio for Various Fuels

| Lambda ($\lambda$) | Gasoline ($C_8H_{17}$) | Methanol ($CH_3OH$) | Diesel ($C_{12}H_{26}$) | E85 (~ only) |
|---|---|---|---|---|
| 0.75 | 11.0:1 | 4.8:1 | 10.9:1 | ~7.28:1 |
| 0.80 | 11.8:1 | 5.1:1 | 11.6:1 | ~7.76:1 |
| 0.85 | 12.5:1 | 5.4:1 | 12.3:1 | ~8.25:1 |
| 0.90 | 13.2:1 | 5.8:1 | 13.7:1 | ~8.73:1 |
| 0.95 | 14.0:1 | 6.0:1 | 13.8:1 | ~9.22:1 |
| 1.00 | 14.7:1 | 6.4:1 | 14.5:1 | ~9.70:1* |
| 1.05 | 15.4:1 | 6.4:1 | 15.2:1 | ~10.19:1 |
| 1.10 | 16.2:1 | 6.7:1 | 16.0:1 | ~10.67:1 |
| 1.15 | 16.9:1 | 7.0:1 | 16.7:1 | ~11.16:1 |
| 1.20 | 17.6:1 | 7.4:1 | 17.4:1 | ~11.64:1 |
| 1.25 | 18.4:1 | 7.7:1 | 18.1:1 | ~12.13:1 |

*The same engine using pump grade E85 (85% ethanol and 15% gasoline) will have an expected A/F ratio of about 9.7 or 9.8:1. However it is very difficult to give a precise estimation for E85 fuel because the normal variations in both the amount of gasoline and the amount of ethanol swing by more than 5%. In Brazil, one of their fuels is Gasohol-22 (ethanol and 22% gasoline) and has a stoichiometric ratio of 13.28:1. The fuel is a mix of gasoline and ethanol. The idea of ethanol and gasoline mixes is not new and has been around for a long time. All that has varied is the price, percentages and availability.

horsepower. But burning fuel produces the energy to make it all happen. The best condition for making horsepower is when the fuel is completely burned for the amount of air that is available. That particular ratio of air to fuel is called the *stoichiometric* or *chemically correct* ratio. For "normal" gasoline, that stoichiometric number is typically 14.7:1. However it is important to point out that there are hundreds of compounds that are sold with all being called "gasoline". The real A/F ratios of various gasolines (if referred to at stoichiometric) range from about 14.5:1 to 15:1. It all depends on the carbon and hydrogen atom content.

*Lambda*—The stoichiometric reference would also be equal to a Lambda ($\lambda$) of 1.00. Lambda is a Greek letter used to reference a standard A/F ratio. Understanding how Lambda works will help you tune an internal combustion engine for any given liquid fuel. Using the Lambda reference is quite often an easier way to identify rich or lean mixtures in tuning.

The expected A/F ratio for normal pump gasoline is approximately 14.7:1. The big question is: What is normal pump gasoline these days? Pump gas might have some percentage of ethanol (8% to 15% is common in some areas). Pump gas is certainly not the same in New Jersey as it is in California, Rhode Island, Montana or Texas. This applies to any of the various states for that matter. It

The internal modifications of ports have the intended purpose of getting more air (thus more oxygen) into the cylinders and more exhaust gases out. A trash can makes an easy workbench for occasional projects and helps with cleanups.

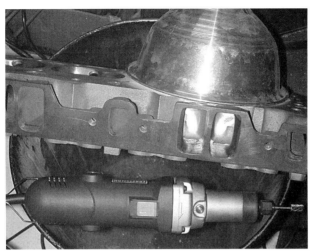

 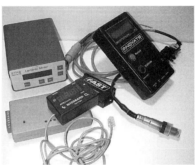

These components and instruments allow a tuner to reference either the A/F ratio or the Lambda by placing a sensor in the exhaust gas stream. All is not as easy as it seems however and the tuner needs to know about response times and relationships of the instrumentation.

This hand-held device is something that many modern tuners cannot do without and often rely on the numbers produced so greatly they are sometimes at a disadvantage if the equipment doesn't work properly or they foul a sensor. You need to know what variables have an effect on the numbers before you go in that direction.

is very difficult to even get exactly the same fuel at several different gas stations. Take a closer look at the section on page 27, *Chemistry of the Fuel*. You need to know precisely what is in the fuel that you plan to use. Do not assume that all pump gas is the same, because it isn't.

Note that this is the reference A/F ratio base and as a result need not be extremely precise as long as the others being compared vary by the correct difference. The result is what is called trend information and is more than adequate for the use of this tool of comparison and tuning. What you want to see is that if you add 10% in fuel, that there is a shift in data that reflects that difference ($\Delta$) from the base. That is one reason why the use of Lambda ($\lambda$) for reference can be so handy, but you must understand how those numbers came to be so you have comfort in knowing how to use them properly.

The reference to Lambda is such that at a Lambda value of 1.00 the mixture is considered to be stoichiometric or chemically correct. Mixtures less than 1.00 are considered rich; mixtures more than a Lambda value of 1.00 are considered lean. Mathematically stated, Lambda is:

$\lambda = A/F \div A/F_{stoich}$

and of course with some algebraic manipulation:

$A/F = A/F_{stoich} \times \lambda$
where A/F = air/fuel ratio and $A/F_{stoich}$ = air/fuel at stoichiometric or chemically correct for complete burn.

Because you might also see something called the "equivalence ratio" used in some articles or textbooks, be aware that:

$\Phi = 1 \div \lambda$, which is the mathematical reciprocal of Lambda ($1/\lambda$)
where $\Phi$ = equivalence ratio (Greek letter Phi), and $\lambda$ is Lambda, defined previously. In other words, the equivalence ratio is inversely proportional to Lambda.

There are any numbers of devices that can be inserted in the exhaust tract to help to identify the A/F ratio or relate to Lambda. Most of the instruments commonly used in EFI (electronic fuel injection) set ups because most EFI units use an $O_2$ sensor in the exhaust to provide feedback to the control system. Utilizing that type of analysis technology on an engine that is equipped with a carburetor or mechanical fuel injection system makes tuning a much more pleasurable experience. Using a sensor in the exhaust located properly can help you get to the proper tune-up in a much

# INTERNAL COMBUSTION ENGINE FUNDAMENTALS

Every one of these plugs has been run in engines and each one tells a bit different story. Just looking at the ends for some general color appearances is only part of the process. In order to look at the details requires a magnifying plug light. In a particular engine the dark plug produced about 10 hp more with excess fuel (RBT) because the engine was prone to detonate. The ground electrode was cut for piston dome clearance.

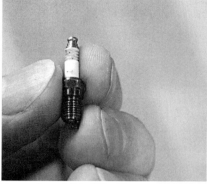

Try and read this little Knapp rimfire spark plug without using a magnifying spark plug light. The threads on this tiny specimen are #10 x 40 per inch. Photo courtesy Conley Precision Engines.

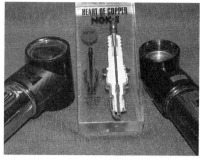

The spark plug light on the left was made by Champion Spark Plug Co. and the one on the right is a lit loupe available from chain stores. The Champion light is better because of the angle of the light bulb so you can see inside the plug down on the insulator and that is where you need to be looking.

shorter timeframe.

Most dynamometer operators today rely strongly on the data from exhaust-mounted sensors. The exhaust sensor (preferably the wideband type) must be mounted in the top 180° of the exhaust system so that water won't affect the sensor. It must also be mounted far enough away from the end of the pipe so that the atmospheric conditions won't affect the readings that the sensor provides.

Because of multiple variables concerning exhaust heat and thermocouple mounting and time responses to changes, it is strongly suggested that you do not use EGTs (exhaust gas thermocouples) for tuning references. However, exhaust gas thermocouples are very good tools providing an indication for looking at such things as intake manifold distribution.

## Reading Spark Plugs and Tuning Tips

Understanding how the heat transfers from the spark plug tip to the cylinder head is also an important thing to know before making any tuning adjustments.

In reality "reading spark plugs" is at best a guessing game that is very dependent upon your experience and knowledge. Most spark plug manufacturers try and discourage putting much faith in attempting to read the spark plugs. However, a few hints will be presented that might help you to ascertain some things about your engine and your tune-up selection.

Given enough experience and practice, the spark plug can give you some valuable information about what is going on in the combustion chamber. Some "tuners" say they can eyeball a plug tip and tell what's going on—which I think is a lot of BS! In order to make any sense of what the plug is telling you, it must be carefully inspected with a good magnifier and light. Everything that you need to see is very small, so the proper light and magnifier is the only way to get the job done.

The tip and end of the spark plug is the easiest thing to look at first. If the ground and center electrodes are still intact, then there is nothing to fear about what is going on in the combustion chamber. However, closer inspection can reveal important details. Most manufacturers say that it takes some time to color the plug sufficiently to really have an indication of what is going on in the combustion chamber, but drag racers often depend on a single pass down the quarter mile or eighth mile racetrack to read the plugs.

Specks on the ceramic insulator nose of the plug can tell you things about the combustion process. If the tiny specks are little balls or flecks of aluminum-like material, they are probably piston particles. Not good. If the specks are tiny specks of pepper looking material, that is also a sure sign of detonation. Not good. Add fuel and perhaps decrease the spark advance some, too, before more damage is done.

Sometimes the ceramic insulator might be cracked and pieces missing. This is a sure sign of danger requiring closer investigation on where the parts went and a change in the tune up.

If the ceramic insulator nose is chalky white has a white, satin-like sheen, this indicates the plug heat range is too hot. If the center electrode has a look of cement boil between it and the ceramic insulator,

# Engine Airflow

This racer has pulled a plug out (after a pass on the dragstrip) to check the plug indications and they are close to being spot on so he should leave the timing and the jetting alone. When this photo was taken, the plug was still hot.

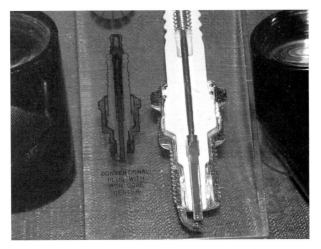

This plug cutaway view helps you to learn how a spark plug is built. They are not all the same, but in general this is the way they are made. Not all manufacturers use ceramic glue around the center electrode, so be aware of that.

this also indicates a heat range that is too hot. Choose a colder heat range. The ideal heat range is one that runs cool enough at high speed and yet doesn't foul at idle. Or the ideal heat range can be thought to be one that won't foul at idle and won't cause detonation because of elevated tip temperatures. Both descriptions are saying essentially the same thing.

The junction of the ceramic insulator and the plug body is where the heat transfers from the plug tip to the cylinder head and then to the coolant. The heat transfer path will show as a defined ring that looks almost like a shadow. That is why you must use a light with the magnifier—so that it will illuminate the area deep down in the plug. That is also why colder plugs are much easier to read—because they are easier to see. When the ring is close to the nose of the ceramic insulator, that indicates the mixture is richer than when the ring is more toward the bottom of the ceramic. Note that you are looking at the plug when it is inverted (upside down).

Take a look at the ground strap that is connected to the threaded portion of the plug body. It should have a clean appearance and be slightly colored (perhaps grey to very light tan) which depends on the dye in the gasoline fuel. The ground strap can indicate the heat right at the tip of the strap and it should not be discolored very far down the strap toward the connection with the plug-threaded portion. If the electrodes are blue from heat, that would indicate a plug tip temperature that is too hot. A colder plug heat range would be of help.

Next take a look at the color indication on the threads. The number of the threads discolored indicates how hot the end of the spark plug has been. Normally, a maximum of three threads of heat indication is all you ever want. Most engines are much "happier" with only two threads colored from heat in the combustion chamber.

Reading plugs on methanol fuel is much more difficult than when burning gasoline for fuel. But it can be done if you approach the process carefully and learn to look through the inspection light magnifier and get focused on the details that you should observe. Methanol burns much cleaner than gasoline does and the plugs will often look (to the uninformed) like they haven't even fired.

Look closely and you will see the telltale marks of combustion on the end of the plug called a "spark mark" and the ground strap will be discolored as it goes toward the threaded portion of the plug. The plug ground strap should not be discolored past the bend in the strap. If it is, add fuel and perhaps decrease the spark advance.

Checking the discoloration of the threads for signs of heat are much the same as when the engine is burning gasoline.

The very best method to apply in order to learn to properly read spark plugs is to run the engine on either a chassis dyno or an engine dyno and tune it for best power. When that is accomplished, take a look at the spark plugs and try to duplicate that at the race track. If the engine is to be run in a distance event or at Bonneville, the mixture had better be on the rich side or you will be going home early and with fewer usable parts than when you first started the engine.

Unfortunately there is way too much advice in the pits to "lean 'er down" rather than "fatten 'er up." Those who are "track savvy" with greater skill and experience would tend to throw more fuel at a combination until it slows down, a much wiser and safer approach.

## Testing for MBT

The best spark advance selection is normally established during engine dynamometer testing and is referred to as MBT (Minimum spark advance for Best Torque) or a spark loop search. There is much

There is a lot more going on in this photo than just timing the engine on a dynamometer. What the operator is looking at is history and the dial type timing light is not a good source for accuracy. Minimum spark advance for the best torque (MBT) is the goal.

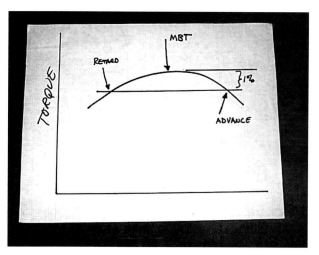

This sketch demonstrates graphically the location for MBT. It is not an easy task on a dynamometer but it is much more difficult trying to find that point on the racetrack. Lack of patience has no place in effective engine development.

more to correct ignition timing than guessing "about 38 degrees or 40 degrees BTDC 'cause that's what everybody does."

It is next to impossible to do this type of testing without having access to a properly instrumented dynamometer. The ignition timing issue is one of the least understood relationships in an engine but it is more important than generally considered. It is far better to make minor adjustments in timing than to guess what advance number to use.

MBT testing is done at steady state (fixed rpm). For impatient testers, some testing can be done at transient rates, but not to exceed 100 rpm/sec as faster acceleration rates will skew the results to apply too much timing.

Testing for MBT using an engine dynamometer is a simple process, but can be time consuming if done at various engine rpm points to establish a complete spark curve. The dynamometer is placed in servo control to target the peak torque rpm point. The most common method of testing for MBT is done at WOT and at the engine's peak torque point. The timing is advanced 2 degrees at a time until the final 2 degrees does not increase the torque. At that point, the timing is retarded 1 degree and is referred to as MBT.

Another method for MBT testing is again referenced to peak torque and the timing is advanced until there is a loss of 1% in torque reading. Then the timing is retarded until there is a loss of 1% in torque reading. Halfway between the two points is MBT. Sometimes the maximum advance point is referred to as MBT + 4 degrees. For a performance application, the MBT point should be very carefully established with good fuel and appropriate caution.

If a tuner wishes to vary the timing and verify the engine's sensitivity to spark advance, then the procedure is the same as above, with the addition of retarding the timing for a point of 3% and 5% or more torque loss, so that those points are known.

An easy method for adjusting the timing should be used from the operations console of a dynamometer. Although many types exist, the most popular method is to use an electronic spark control such as the Autotronic Controls MSD 8680 for batch timing control or the MSD 7553 dyno tuning programmer. The 7553 provides individual cylinder timing that can be the guideline for making mechanical changes in the reluctor on magnetic triggered systems.

Some electronically controlled engines can be adjusted by communicating with the ECU via an external controller. This method is by far the easiest approach, but the process can also be done mechanically, it just takes more time in testing.

Selecting the correct timing light is almost as critical as the MBT process itself. There are only a few timing lights that are recommended for use on high-performance engines. The one made by Sears and the ones made by MSD are not the most expensive timing lights available, but they are the most accurate. Many timing lights suffer from a phenomenon called "inductive retard," with the Sears and MSD units having less than a few tenths of a degree of timing fault. On the other hand you can spend megabucks for a dial-your-own timing light and get much less accuracy (such as up to 2 degrees per thousand rpm). So what appears to be 38 degrees BTDC at 5000 rpm, the real number might be something like 48 degrees BTDC!

# Chapter 3
# Engine Airflow Relationships

*Today's scientists have substituted mathematics for experiments, and they wander off through equation after equation, and eventually build a structure which has no relation to reality.* —Nikola Tesla

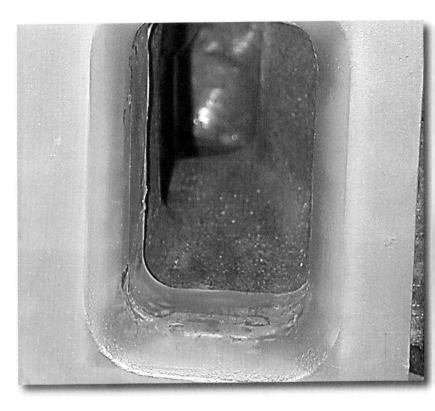

How much air will flow through this intake port? Typical answers are provided by measurement on an airflow bench. The bench flow numbers can relate how a specific engine will perform.

There are many different internal combustion engine designs, but in this book we will primarily concentrate on those that use pistons, connecting rods, and crankshafts. These type engines are referred to as reciprocating engines—the pistons go up and down and the crankshaft goes round and round.

## Cubic Feet Per Minute (cfm)

*Cubic feet per minute* refers to the airflow of either components when measured on a flow bench, or how much air the engine consumes. This reference is for volume flow and the airflow measurement will be one of the central figures used when studying airflow through engines. The capability of the engine to produce horsepower is directly proportional to its capacity to flow air. Normal engines consume about 1.5 cfm (or so) per horsepower.

## Volumetric Efficiency (VE)

*Volumetric Efficiency (VE)* is a reference to how well the engine uses the air that is available, expressed as a percentage. In reality, the term volumetric efficiency is actually a misnomer because it is really related to mass and not volume. Another term that is closely related is the term *brake specific air consumption (BSAC)* which refers to how efficiently the engine used the air that it consumed in lb/hp-hr (pounds per horsepower hour). The formula for VE is:

**VE% = $\eta_v$ = (ma ÷ $\rho_{ai}$ x $V_d$)**
where $\eta_v$ = Greek letter Eta, v = volumetric efficiency, ma = mass of air, $\rho_{ai}$ = Greek letter rho (air inlet density), $V_d$ = volume displaced. Note that in this classical form, the air density and not just the volume is the reference.

Most dynamometer data acquisition systems do not calculate the VE percentage in exactly the same way as listed above. Most of the dynamometer instrumentation systems use a much different approach. Some dynamometer manufacturers use this equation or something similar:

**VE% = (cfm x 3456 x 100%) ÷ (rpm x displacement)**
where VE% = volumetric efficiency, cfm = airflow in cubic feet per minute, rpm = engine revolutions per minute, displacement = cubic inches for complete engine

The VE of a typical engine is usually between 80% and 90%. It is not uncommon for dynamometer data sheets to list VE at between 95% and 105% for a naturally aspirated engine. Flathead type (valve in block) engines will typically only have 65% to 75% volumetric efficiency, even when modified. Whereas a four-valve pent-roof combustion

# Engine Airflow Relationships

These devices are air turbines that measure the airflow into the engine when placed on the carburetor or throttle body when the engine is on a dyno. The readout on a display will show the cfm of air the engine is passing and provide other related information. Photo courtesy Land and Sea Dynamometers, Inc.

Using a Pitot tube with a flow bench can give you information on the velocity of the airflow inside the port. However you need to learn how to use the Pitot tube properly so you'll be able to interpret the collected information. For details on Pitot tubes, see page 64 in Chapter 5.

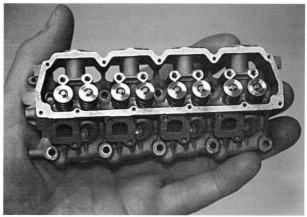

All components flowing air are important no matter what size the engine or cylinder heads. This cylinder head is really part of a V8 engine. Either it is really small or the hand holding it is huge. A gearhead dentist could port these little heads very nicely. Photo courtesy Conley Precision Engines.

## How Much Air Does an Engine Use?

The exact number is a direct function of how efficiently the engine can use the airflow. It is best to start off with a discussion of naturally aspirated engines.

An efficient, well-tuned internal combustion four-stroke cycle piston engine running gasoline will use about 1.25 cfm/hp at peak torque and about 1.4 cfm/hp at peak power at wide open throttle. If the engine uses more than these numbers, it is not running very efficiently. However, at high engine speeds (rpm) and using camshafts of high overlap, the cfm/hp number might end up being greater by 10% to 15% or so. Good design parameters would indicate that the *available* airflow should be in excess of the numbers listed above by at least 20% (1.5 and 1.68 cfm/hp, respectively) and perhaps more in some applications.

If an engine does not seal very well, it would use much more air. The same goes for a condition that might exist when an engine has the incorrect camshaft and valve timing. On an engine dyno, the term BSAC (brake specific air consumption) is used

chamber engine (normally overhead cam heads) can see as much as 115% or more volumetric efficiency at very high engine speeds (as in 20,000 rpm). The high engine speed applications are greatly dependent upon a phenomenon known as *inertia supercharging* which is presented in the section on intake and exhaust tuning, page 106.

If the VE is high and the BSAC (brake specific air consumption) is also high, the air and fuel might be getting tossed out in the exhaust stream and this data could be trying to tell you to spread the LCA (lobe center angle also called the lobe separation angle which is a peak lift to peak lift reference) on the camshaft in order to better capture the flow.

You need to find out where all the specific numbers come from for them to mean something you can relate to.

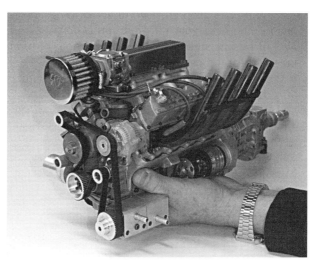

Sort of puts things in perspective doesn't it? This engine weighs about 11 lb. and produces in excess of 6.5 horsepower at 9500 rpm. The engine has a displacement of 6.09 cubic inches and is a self-driven air pump. Photo courtesy Conley Precision Engines.

39

# Engine Airflow

**This cylinder head is a modified Chevy Super Stock that flows a bit over 260 cfm at 28 inches test pressure. That is impressive because the intake valve is 1.94" diameter. The cylinder head airflow capacity greatly affects the engine's volumetric efficiency and its horsepower.**

to refer to how efficiently the engine turned air into power. If the engine is consuming lots of air and not making high levels of specific power, it could be tossing away the fuel and air right out the exhaust during the overlap event. The comparison of BSAC and volumetric efficiency (VE) helps you to see how well the camshaft fits the engine build combination.

The calculation of the amount of airflow in cfm that an engine uses is pretty straightforward. An engine has a specific volume, called the *swept volume*, which is the displacement as it moves through two full revolutions (720 degrees of crankshaft rotation). Because we would like to have the answer in cubic feet per minute (cfm), you need to convert the swept volume from cubic inches to cubic feet. There are 1728 cubic inches in 1 cubic foot (12 inches x 12 inches x 12 inches = 1728 cubic inches). As an example, consider an engine with 302 cubic inches displacement at 6500 rpm:

cubic feet = 302 ÷ 1728 = 0.1748 cubic feet
6500 rpm ÷ 2 = 3250 rpm
cfm = 0.1748 cubic feet x 3250 rpm = 568 cfm

Remember that is based on 100% volumetric efficiency. If you want to check out other reference volumetric efficiencies, just multiply the answer above by the decimal equivalent of the percentage, such as 80% = 0.8, 75% = 0.75 and so forth.

Now if everything was nice and rosy and the numbers listed for the amount of airflow needed for power at peak power rpm were true, then the following quick evaluation would be close:

568 cfm ÷ 1.4 cfm/hp = 405.7hp at 6500 rpm

You might have seen some similar references to this type of analysis of engine airflow if you have ever gathered information on trying to figure out how to size carburetors. If so, you probably saw it written as:

cfm = (CID ÷ 2 x rpm ÷ 1728) x VE%

which simplifies to:

cfm = (CID x rpm ÷ 3456) x VE%
where cfm = cubic feet per minute (airflow), CID = cubic inches displacement, rpm = revolutions per minute, VE% = volumetric efficiency. **The constant 3456 comes from the fact that there are 1728 cubic inches per cubic foot, and the engine takes two revolutions to complete all four cycles (1728 x 2 = 3456). Now you know where the numbers come from.**

Using the engine size and rpm in our example listed previously will give exactly the same answer of 568 cfm (rounded off from 567.99). Note however that this number will change if you use a VE% number that is more or less than 100%.

Unless it is presented differently, most of the examples in this book will assume a target VE of 100%. That way it makes the math a lot easier and more fun at the same time. Now technically speaking that is not exactly correct, but for most of our work using 100% volumetric efficiency is just fine. By this definition, VE is related to the swept volume of the engine.

However it is necessary to note that these flow numbers do not have a test pressure or differential pressure (ΔP) for a reference. And you will find stated several places in this book that flow numbers without nowing the ΔP at which the data was collected can be a meaningless number.

Another way to look at how much air an engine uses is by looking at the ratio of air to fuel. Because that power is the result of burning fuel efficiently, an appropriate amount of atmospheric air must also accompany the fuel, thus the air/fuel ratio or A/F reference. If we look at the fuel requirement for making only 100 horsepower for just 10 seconds, the amount of airflow will follow the A/F ratio accordingly.

If the BSFC (brake specific fuel consumption) is approximately 0.5 lb/hp-hr for gasoline, then 100 hp would require a fuel flow of at least 50 lb/hr or 50 lb/hr divided by 3,600 seconds = 0.01389 lb/sec. So, in 10 seconds the fuel consumed would be:

10 x 0.01389 = 0.1389 lb.

# ENGINE AIRFLOW RELATIONSHIPS

Never underestimate an opponent. Restrictions in the engine need to be decreased and the friction needs to be diminished if possible. Even something as simple as getting cool air to the engine intake is worthwhile to investigate because it is free power. Smart technicians know this already. That is why they call it "beat the heat."

Maximum power is typically produced on gasoline at an A/F ratio of about 12.5:1, and 12.5 x 0.1389 = 1.7363 lb. of air would be expected to go along with the fuel.
At sea level conditions (29.92 in.Hg, 60°F) a cubic foot of air weighs 0.0763 lb. For those conditions, 1.7363 ÷ 0.0763 = 22.7562 cubic feet of air.

Now let's take the previous process and evaluate the conditions for producing 500 hp for the same timeframe of 10 seconds. This would be something equivalent to a drag car of about 130 mph weighing approximately 3,000 lb. The fuel used in such a pass would be:

500 hp x 0.5 lb/hp-hr = 250 lb/hr
250 lb/hr ÷ 3,600 seconds = 0.06944 lb/sec
0.06944 lb/sec x 10 sec = 0.6944 lb. of fuel, which would be slightly more than a tenth of a gallon of fuel used.

How much air at an A/F ratio of 12.5:1?

12.5 x 0.6944 = 8.68 lb. of air.
8.68 ÷ 0.0763 = 113.76 cubic feet of air in 10 seconds or 113.76 ÷ 10 = 11.376 cubic feet per second, which is 682.58 cfm (11.376 cubic feet ÷ sec x 60 sec/min = 682.58 cubic feet/min or cfm).
8.68 lb/sec x 60 sec/min = 520.8 lb/min. A pretty good rule of thumb is that an airflow rate (mass) of 10 lb/min is good for approximately 100 hp.
520.8 ÷ 10 = 52.08 (520 hp).

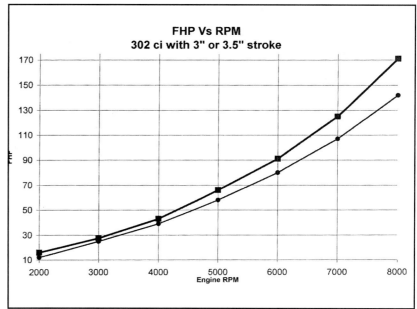

The graph shows how varying the stroke from 3" to 3.5" causes an increase in the internal friction power. FHP is literally how much power is required to whirl the crank and related parts. Less internal power loss is better as nothing is for free. Some like to say that this doesn't even exist because it is just too much to think about. Thinking is power.

That is pretty close to the old rule of thumb. One does not look for absolutes here, just good approximations, so this becomes another handy estimation tool in your trick bag.

Note that some engines are more efficient and BSFC numbers at peak power rpm of 0.45 lb/hp-hr to 0.47 lb/hp-hr are not uncommon with well-developed cylinder heads and complementary components. At peak torque, the BSFC numbers typically become even more efficient and a BSFC number of 0.38 lb/hp-hr is not uncommon on a well-developed engine.

You see how all this can come together to make some sense of the airflow and numbers? It all must be in conjunction with how much fuel we can get the engine to burn if we truly want to make power instead of just another noisy and smelly machine. Is this stuff fun or what?

## Friction Horsepower

Because of the internal airflow use of the engine, consider that whirling all that metal must take some power to do. The FHP or friction horsepower varies with rpm and stroke more than it varies with piston area. Pretend for a moment that an engine was running and it was not connected to any load or requirement to do work. This is best thought of as an engine running in neutral gear. Does it use air? Does it use fuel? Of course it does. If you were to open the throttle a bit and the engine rpm increased as a result, would it use more fuel? The answer is again, yes. What happens to the power that is the result of burning fuel and air? As a matter of explanation, it provides the required power to whirl all that metal. Well, how much power would that be?

That is where it gets interesting. If you attached an air measuring device to the same engine in the previous example and recorded the airflow at each rpm data point (such as every one thousand rpm), then you would find the following.

At 1000 rpm the engine uses less airflow than it does at 3000 rpm and it uses a lot more at 6000 rpm. If you had those airflow numbers at each of the rpm data points (every 1000 rpm from 1000–6000 rpm) and you divided the airflow number (cfm) by 1.2, you would have the data to draw a curve of the friction power for that particular engine. And yet there are folks out there that don't believe in the term friction horsepower! There are many phenomena that exist within the engine whether you can see them or not.

There's little argument, however, that decreasing the internal friction in the engine is beneficial. Decreasing windage loads on the crankshaft, ring friction and valvetrain loads will pay off in better levels of power without tossing more fuel at the process. However, fuel use is not typically way up on the "fix it" list in the search for more power. Think toward more efficiency and you will gain from your efforts.

If you think of the size of the counterweights on a typical V8 crankshaft as being about 6" in diameter, and the engine at 6000 rpm, there is a speed of the surface of the counterweights that might be of interest when discussing windage. The circumference of the counterweights is approximately:

π x 6" = 18.85" ÷ 12 = 1.57 feet per revolution
1.57 feet x 6000 rpm = 9420 feet per minute.
Since 60 mph = 5280 feet per minute, 9420 feet per minute ÷ 5280 feet per minute =
1.784 x 60 = 107 mph.

And that is only at 6000 rpm. Think of how much resistance is being provided from rotating the crank through the air. Now add in a bit of attached oil and the answer becomes obvious. There is internal windage and it is just a part of the internal loss called friction horsepower (FHP). Or if you are still not a believer, just stick your hand out the window of a car that is traveling down the highway at 60 mph and "feel" the resistance. There is more to scrapers and windage reduction than meets the

### Estimating Friction Power

| $P_s$ ft/min | $F_t$ factor | $P_s$ ft/min | $F_t$ factor |
|---|---|---|---|
| 1,000 | 0.120 | 4,000 | 0.308 |
| 1,500 | 0.145 | 4,500 | 0.368 |
| 2,000 | 0.170 | 5,000 | 0.433 |
| 2,500 | 0.203 | 5,500 | 0.503 |
| 3,000 | 0.233 | 6,000 | 0.578 |
| 3,500 | 0.263 | 6,500 | 0.653 |

By first calculating $P_s$, you can then interpolate for factor values not listed.

eye. It takes careful thought, hard work and hand fitting. A rough estimate of friction power required is presented in the following equations:

FHP = ($F_t$ x rpm) ÷ 5252
where FHP = Friction Horsepower, $F_t$ = Friction torque in lb-ft, rpm = engine speed in revolutions per minute

$F_t$ = (factor) x CID
where $F_t$ = Friction torque in lb-ft, factor is from the table nearby, CID = displacement of engine in cubic inches. The factor comes from a reference to piston speed and it is found by:

$P_s$ = (S x rpm) ÷ 6
where Ps = piston speed in feet per minute, S = stroke in inches, rpm = engine revolutions per minute

As an example, let's find the internal friction power required to run a 377 cubic inch engine to 7500 rpm. The stroke used is 3.48". First you must calculate piston speed. The equation for piston speed is:

$P_s$ = (S x rpm) ÷ 6
So $P_s$ = (3.48 x 7500) ÷ 6 = 4350 feet per minute (ft/min).

As you can see from the table above, the piston speed falls between 4000 and 4500, so the reference factor would be 0.350. Then take the factor and multiply it times the piston speed and by the engine's displacement.

Ft = (0.350) x 377 = 131.95 lb-ft
FHP = (Ft x rpm) ÷ 5252
= (131.95 x 7500) ÷ 5252
= 188.43 FHP

This is just the amount needed to whirl all that

iron and aluminum around inside the engine.

Although that seems amazing, it really is pretty logical when you think about it. If you want to do the rest of the calculations for an estimated friction power curve for the described engine, have at it. Start at about 2500 rpm and calculate the answer at each 500 rpm.

## Manifold Pressures

About how much pressure is in the manifold at wide open throttle (WOT)? Well, the number varies and is directly related to the engine rpm and the size of the carburetor or the throttle body.

For this discussion, we will assume that the carburetor is properly sized (not too big or too small). The manifold vacuum at WOT at peak engine rpm might be something close to 1 in.Hg (inches of mercury) or a bit less. 1 in.Hg = 13.6 in.$H_2O$ (inches of water), and that number is an average reading in the manifold. Even with the largest carburetor setup the engine should have something close to at least 6 or 7" of water (about 0.5 in.Hg) so that the carburetor will have a usable main booster signal. The actual differential pressure ($\Delta P$) average for the manifold will be varied with valve opening and other issues such as rpm, but a very good number for estimates would be between 1 in.Hg and 3 in.Hg, depending upon where in the manifold, plenum, or port the pressure was measured. So, if one takes a quick and dirty average of the previous information and used 2 in.Hg, which is equal to 27.2 in.$H_2O$, then the "almost" or quasi industry standard number of 28 in.$H_2O$ makes a little more sense.

Although there is not really an industry standard for test pressure, several points about this issue will be presented later and then a standard and some suggested testing procedures will be proposed. The historical reasoning that 28 in.$H_2O$ was used as a reference is somewhat interesting because some early flow bench testing was done at up to 68 in.$H_2O$. It seems as if the test pressure issue was not very important at first and then as more and more engineers and gearheads got involved in both testing and building flow benches it became one more thing that was common to several test sites.

**Knowing what goes on in the intake manifold will help you make more power. This Ford engine makes more than 700 hp and does not happen by accident. Careful attention to all the details in the flow path helps. Photo courtesy Cathy Bevers.**

# Chapter 4
# How Flow Benches Work

*Those who cannot learn from history are doomed to repeat it.* —George Santayana

These components are for my first home-built flow bench from the early 1970s. The airflow source was a shop vacuum cleaner. The digital output provided a capacity of 350 cfm and still works. Earlier exposures in college fluid labs were where an LFE was used. Missing in this photo is the U-tube manometer for test pressures. This was advanced stuff for that time frame.

Contrary to what some would have you believe, a flow bench is a pretty simple device and the technology on their inner workings has been around for well over a hundred years. The primary description of a flow bench is that it is a machine that measures the resistance to flow air through any component. Flow benches are based on some very straightforward, basic principles of airflow.

## A Brief History of Flow Benches and Airflow Development

The first device to study airflow in the United States that we could easily relate to was the small wind tunnel built in 1896 by the Wright brothers to study wing designs. Most modern flow benches provide more power to the air than the piece that the Wright brothers produced. The Wright brothers' tunnel was powered by a one horsepower engine and was supposedly capable of about 25 to 30 mph wind velocity. Today in fact, it is not uncommon for some shops to have 20 to 25 horsepower available for airflow testing cylinder heads and manifolds and regularly produce air speeds of 80+mph on large cross-sections and up to 200+mph on smaller cross-sections.

Some of the first airflow studies that were done specifically for engine development date back to an early paper done by the aircraft engine manufacturers (1905–1908). Valves, cylinder heads, and manifolds were investigated for the purpose of a need for more horsepower and engine efficiency. The first airflow benches used at the OEM (original equipment manufacturer) level were expensive, cumbersome and complex machines. Oldsmobile and Pontiac used flow bench-guided designs early on. However, Chevrolet did not use a flow bench lab until the 1970s. American Motors used flow bench-guided cylinder head designs in the early 1970s. Chrysler adapted a flow lab from elements that were used in air filter work and the lab was developed in parallel with their introduction of the infamous 426 Hemi engine. The Ford Motor Company flow lab dates to the mid-1960s, when they used it to develop their LeMans-winning GT40 race cars. The Ford testing and development program was significant, in that they adapted equipment to build a flow stand for cylinder head and manifold development. They were flow testing at 5 in.Hg, which equates to 68 in.$H_2O$ test pressure.

Many of the OEMs have abandoned their in-house airflow benches in favor of outsourcing to any number of commercial airflow testing companies, because it is more cost-effective. The combination of high-level engineering knowledge and hands-on development work speeds the process of developing parts.

The same OEMs have concentrated a great deal on

# How Flow Benches Work

This is currently my favorite all around flow bench. It is the D-680 from Saenz and Jamison Equipment. The D-680 has a digital screen to read data on for test pressure and flow. It has the capacity of 680 cfm at 28" of water. You can do much with this bench including adding a computer as shown.

Any shop that is serious about their quality control and wants to improve their work product either has or has ready access to a flow bench. This shop has a 600 cfm bench that is rated at 28 in.$H_2O$ test pressure. They use it on a regular basis. Bore adapter is in place waiting on a cylinder head to be mounted.

improving their use of CFD (computational fluid dynamics) programs in order to improve the timeline on airflow development. Because the CFD time is very expensive (as are the programs computer programs), many CFD development programs are still using shops to do the initial flow testing.

Although flow benches and airflow data have been part of the internal combustion engine development cycle for design, research and development for many years, the detailed study of and relationship to performance has only been common in the racing industry for about the last 35 to 40 years.

The foundry process and the associated compromises actually controlled most early cylinder head and manifold designs. These manufacturing compromises drove most designs, not the technical aspects or specific airflow requirements.

Today there are thousands of flow benches used around the world every day, and as a result, engine airflow technology continues to develop. There are even many forums and chat rooms on the Internet where tremendous amounts of technical information are openly exchanged. The opinions of what the data means are very diverse, ranging from old wives' tales and questionable theories, to those founded on sound engineering principles and practical application of physical laws.

*Early Airflow Research Pioneers*—Warren Brownfield of Airflow Research used a flow bench early in the development of his cylinder heads and offered a smaller version of it for sale in 1970. They were priced at about $2,000 each, which put off a lot of people. Then Guy Williams from Tulsa, Oklahoma, introduced a portable and more affordable flow bench in 1972, which made it available to many engine builders who couldn't afford one previously. The first model was a tabletop unit and sold for about $800. The same model today sells for substantially more in the United States (over $3,000 at this writing) and considerably more in other countries. But this unit made it possible for almost anyone to take measurements to compare and rate components for the engine.

C.R. Axtell built some laminar flow element (LFE) type (using Meriam units) benches for a few people in the late 1960s and early 1970s, but they were large and bulky and required a large amount of floor space. Jerry Branch (Branch Flowmetrics) had one on the West Coast that worked some with the magazines, and the Edelbrock Corporation (when Edelbrock R&D was under the direction of Jim McFarland) built their own bench with Axtell's assistance around the same time (prior to 1969).

Bob Mullen an ex-Chrysler engineer (one of the original Ramchargers) and cylinder head airflow expert, built his own flow bench. It operated at about 3–4 in.$H_2O$. Mullen was also the chief engineer at Donovan Engineering, where he became know for his airflow work while testing and developing their products in 1971.

The late, legendary Smokey Yunick was also an early user and proponent of the flow bench. He built one somewhere around the late 1950s or early

# ENGINE AIRFLOW

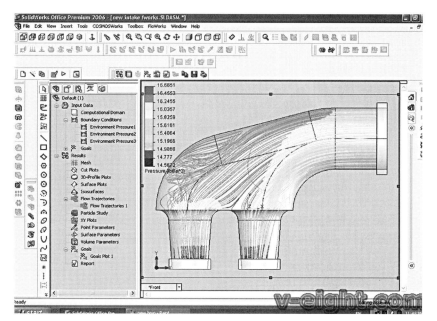

The use of CFD has gained popularity in recent years but is very expensive and needs a very fast computer to work properly. Some of these programs cost $50,000 per seat. Various colors would show different velocities. Flow benches still seem more cost effective. Photo courtesy V-eight.com.

Legendary head porter Joe Mondello is directing a student to work in a specific area of an H-D head. Joe's porting school is one of very few that teaches this skill. Photo courtesy Mondello Technical School.

To keep the cast iron or aluminum cuttings out of the lungs a breathing mask should be worn. Eye and ear protection are also necessary for the safety conscious. It is not a requirement to have a super porting bench, but it is nice if room is available.

1960s to develop his championship-winning stock car engines. Smokey's bench also used a laminar flow element and instrumentation from the Meriam Company (today called Meriam Process Technologies).

Ken Sperry was in charge of the flow lab for CPC (Chevrolet, Pontiac, Canada Group) for years. Ken's efforts standardized many of the test procedures that were followed by both Chevrolet and its many contractors doing engine airflow work. The test bench that Ken built was powered by an electric motor-driven GMC 6-71 positive displacement blower and used a laminar flow element (LFE) for accurate flow measurements. A GMC 6-71 blower uses two three-lobed helically twisted rotors that are meshed and geared together. The blower moves approximately 411 cubic inches of air per revolution, depending on the condition and efficiency of the blower. 1728 cubic inches are in a cubic foot, so 411 ÷ 1728 = 0.23785 cubic feet per revolution. The test rig used a variable drive 10 hp motor, so the cfm capacity could be varied with the rpm of the drive mechanism.

The various and multiple test pressures that used in the field are interesting in that they were probably derived as a historical reference more than a technical requirement. This issue will be covered later in the discussion of flow bench data comparison at different test pressures.

In the earlier days of the development of the motorsports aftermarket there were very few places that were full service cylinder head shops. Warren Brownfield formed Airflow Research and was an early leader in airflow development. Larry Ofria's Valley Head Service sprouted up along with Cylinder Heads America and Cylinder Heads West.

One of the oldest full service shops was Heads by Joe Mondello. The legendary Mondello has been working at improving airflow through engines since the days of the flathead Fords and Mercurys. In the San Francisco, California area, Lockerman's Porting Service was a favorite of Can Am race engine builders and many early drag racers.

On the East Coast, Brandywine Cylinder Heads was a notable early pioneer, as was Speedwin Automotive in New York and Tony Fiel, who's heads were marketed through Moroso. Harvey Crane at Crane Cams became involved in the cylinder head business. Racing Head Service was founded in Memphis, Tennessee and later became part of Competition Cams. Butch Elkins, of

# How Flow Benches Work

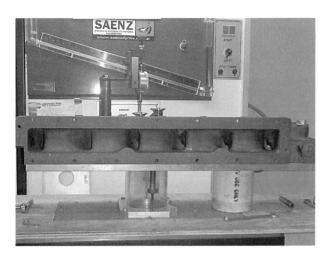

A heavy Cummins diesel head is mounted on the bench here for some study of where the air goes in the problem ports. The device off to the right is just a support column made of PVC. Bench is a special 600 cfm unit. Some interesting things happen when the stock swirl inducing ports are changed. Photo courtesy Lawrence Engines.

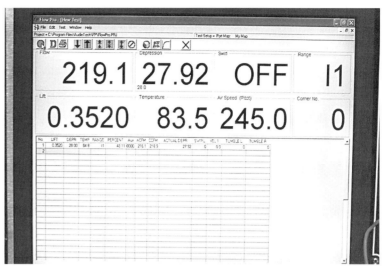

This shows how easy it is to be spoiled by technology. Displaying 219 cfm at a velocity of 245 feet per second and at only 0.352" valve lift, the data is automatically captured on this computer-controlled bench. Bench is rated at 1000 cfm at 28 in.H$_2$0 of test pressure; even the valve lift is automatic. Photo courtesy Ranken Technical College.

Diamond Racing Engines in Detroit, did alot of work with the OEMs, as did C. J. Batten Heads.

Racing spurred a lot of head development. Jim Bell in Texas was a porting force also. The late Lee Shepherd, also from Texas, became nationally recognized for his contributions and was an influential consultant for GM racing for many years, as was his successful Pro Stock racing Reher-Morrison-Shepherd team.

There were many more individuals that helped develop airflow technology that we take for granted today. All of them should be recognized. Remember initially they all did the work by hand because it was long before CNC machines were adapted to cylinder head work. The cylinder head airflow developments of today are the result of some of the most dedicated gearheads that ever picked up a porting tool or did a valve job.

The most successful of these folks did their development with the aid of a flow bench. Much of that assistance came from Edelbrock's research and development group headed by Jim McFarland. During those early years, the OEMs constantly farmed out much of their cylinder head and manifold development to these shops. Gearheads got the job done faster and more efficiently than their OEM engineer counterparts.

The period of time from the late 1950s to the late 1990s has been the most productive in the area of cylinder head designs and modification. As a result, today's the engines are more efficient and more powerful. There is a broader selection of all types of aftermarket cylinder heads available than at any other time in automotive history.

## Types of Flow Benches

There are many kinds of flow bench designs and not all work exactly the same. Consider as a simple example using a shop vacuum cleaner to pull on the exhaust port. As you open the valve, some amount of flow goes past the cylinder and combustion chamber, through the exhaust valve, into the vacuum cleaner and out into the atmosphere. If you had some differential pressure readings, then you would have a very simple flow bench. All flow benches are based upon this basic concept.

*Pitot Tube*—The Pitot tube was named for its inventor Henri Pitot (1695–1771) in 1732. Pitot's first tubes were made out of glass tubing to measure the velocities in the Seine River (France). Think of how much trouble that would cause today because of the MPI (metal particulate index) of modern day wastewater and effluents that probably did not exist when Pitot did his early work. Henri Pitot lived at a time when many others Daniel Bernoulli (1695–1782), Leonhard Euler (1707–1783), Jean de Rond d'Alembert (1717–1783) and himself would much later become recognized as historical icons in the Fluid Mechanics Hall of Fame. Others such as the famous Navier-Stokes equations came later (1822). Claude Louis Navier (1785–1836) and Sir George Gabriel Stokes (1819–1903) followed Bernoulli's initial work and fully applied Newton's second law and the rest is fluid flow history.

The Pitot tube has been described as a stagnation—static tube as it allows measurement of the ratio of stagnation pressure vs. static pressure in

# Engine Airflow

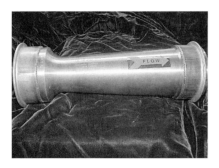 

This critical flow venturi device is shown in profile and at the airflow entry. The pressure differential must be measured accurately and calculations done in order to determine the airflow under the conditions of testing. The expensive unit is stainless steel and is sometimes called a sonic nozzle.

order to calculate velocity. A better description is that the measurement of velocity is by converting the kinetic energy into potential energy at the stagnation point.

The Pitot tube provides a way to probe the port and to supply local velocity numbers. A Pitot tube can be used with a pressure manometer or pressure transducer and special port software to plot areas of activity in the airstream in a port. Pitot tubes are difficult to use in very small ports. Different shapes are needed for intake and exhaust studies. The Pitot tube must face the direction of flow with the end of the tube. The two ports used on the Pitot tube are called the impact port and the static port. The design of the Pitot tube that is used for intake study is in the shape of a hook and the design used for exhaust study is either straight or at 90° to the main shaft of the Pitot tube. Pitot tubes are tricky to use and are sensitive to yaw angles in application. You can even use a Pitot tube to indicate the speed of your vehicle because this is the normal way that aircraft have a way to reference their speed, too. For flow testing using Pitot tubes, see page 64.

*Orifice Flow Benches*—Giovanni Venturi used flat plate orifices for flow measurement as early as 1779, so the concept is certainly not new. However the calculations that can be done today are much more accurate than Venturi's early work.

Orifice flow benches are the most common type, and if calibrated correctly, provide very dependable results across a wide range if multiple orifices are used. These type benches are based on the measurement of a pressure drop across a known and calibrated orifice. Depending on the range of measurement, it might take several orifices to accomplish the job. The orifice type benches are the easiest to verify if there is a question of the calibration over time. They are not typically sensitive to atmospheric conditions and are very tolerant of shop conditions where there might be airborne contaminants that would go through the bench. The orifices used are typically flat plate or sharp edged. Because the output of data is not linear across the full range of any given orifice, as flow demands change, the size of the orifice needs to change (i.e. multiple orifices for a broader range of measurement). This type airflow bench typically uses a vertical liquid manometer for test pressure and an inclined manometer to supply a flow pressure measurement.

*Laminar Flow Element*—The laminar flow element (LFE) benches are not the most complex of the bench designs; however they have some very good features. They are inherently very accurate, but are sensitive to airborne contaminants so they should be equipped with adequate air filters on the LFE. The LFE units cannot be field calibrated, but are very accurate if the proper mathematics are applied by the user/operator. The use of a single LFE allows the operator to use the same measurement device in place throughout a test without the necessity of changing size of the LFE. If the size (capacity) of the LFE is selected properly, a very wide range of flows can be measured with very good precision, but instrumentation needs to be sensitive enough to do the job properly. This type of flow bench uses a vertical liquid manometer for test pressure and an inclined manometer that measures airflow through the LFE.

*Critical Flow Venturi*—As the name implies, this type of flow bench uses specially sized critical airflow venturi. These units are very sensitive to pressure readings and require excellent manometers or very sensitive pressure transducers for pressure drop data. These specially machined venturis are very expensive and require very sensitive instrumentation to achieve good test results. The sizing of the venturi throats are directed to a specific range of flows.

*Hot Wire Anemometer*—These type of benches use modern day electronics to supply data on airflow. They are very sensitive to upstream and downstream vortices and to temperature changes. Thick film and thin film type devices also are in this general category. Many of the electronic fuel injection (EFI) engine management systems use these devices to supply the ECU (electronic control unit) with airflow data. Most of the algorithms used employ multiple points and a great deal of averaging to help dampen rapid changes in data as a

# How Flow Benches Work

I built and used this turbine-type flow bench long ago and the digital readout was quite a feature for the day. The turbine was part of a special airflow measuring system used in an early ultrasonic fuel injection system by Autotronic Controls Corp.

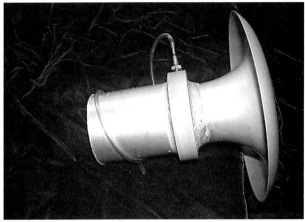

This photo shows an ASME nozzle that has long been the accepted standard for measuring airflow in the turbocharger and aerospace industry. The pressure differential is measured and then the airflow though the device is calculated. A bit cumbersome but accurate and not cheap either.

result of the upstream and downstream vortices that normally occur. It is quite possible with modern electronics and using OEM components for referencing airflow that a bench could be built that would be as accurate and repeatable as OEM automotive EFI and engine ECU controls. This method of measuring the airflow might be an easy process for you to apply if you wanted to do so.

*Turbine or Propeller*—This type of bench uses an enclosed propeller or a turbine in a contained housing as the basis for supplying airflow data. The propeller (turbine) type requires careful calibration for frequency versus cfm. The best packages use a propeller that is of very low mass so that they have low inertia (resistance to motion). The output of these type devices needs electronics to count the frequency of the propeller and also make any corrections for temperature and pressure if required. These type of benches are not very popular, but they certainly have their applications.

## Flow Bench Operation

Airflow benches and flow measurement systems operate on multiple principles and applications of airflow measurement, with the most common being the use of flat plate, sharp-edged orifices or Pitot tube systems. Some designs also use a laminar flow element for measurements. Some designs even use positive displacement measuring devices. Also used for flow measurement are sonic nozzles, ASME (American Society of Mechanical Engineers) nozzles, hot-wire and hot-film anemometers, and laser Doppler anemometers (LDA). The Doppler effect refers to the work of Christian Johann Doppler (1803–1853), an Austrian physicist.

While various problems might be inherent with each design, good results can be achieved if good test procedures are applied and there is a clear path of where the numbers come from.

Regardless of the process applied, typically there is a measurement across a known section (flow pressure), which is the flow measurement and a pressure measurement across the device to be tested (test pressure).

## DO YOU REALLY WANT TO BUILD YOUR OWN FLOW BENCH?

Many people start out their study of airflow by building a homebuilt flow bench. There is nothing wrong with that if you have the time and the talent to put something together. However, do not deceive yourself into thinking that you can save lots of money if you just start out very simply, because it will take lots of your time and energy. Now that the warning has been given, let's take a look at some do-it-yourself options.

It would be best to look around a bit and make sure that you have a very good understanding of the type of bench that you want to build and to also understand how that design works. All those things need to be addressed before you begin to assemble any parts or pieces with which to construct a bench. It might also be a good time to look into the pricing of used homebuilt and commercially built flow benches too. There are all sorts of plans and parts for sale on the Internet, so take a look at those resources as well.

What type of bench to build? The easiest of all is to consider the orifice type benches. They are much easier to construct and the measurement and calibration processes are also much more basic and easy to apply.

What capacity in airflow? If you are a DIY enthusiast that is interested in working with small single-cylinder engines, no need to consider a bench that would be more comfortable working with 1,000 hp engine components. The capacity issue is more along the line of how much airflow power you need. The answer to these questions also pretty much begins to dictate how much floor space you might have to give up. Yes, you can even build a bench to work with model aircraft or marine applications, either two- or four-stroke engines.

Ken Weber of 10 Litre Performance, is leaning on a flow bench he designed and built. It uses a laminar flow element (LFE) as a flow standard. This bench is one of several that Ken has built. His latest version uses a huge LFE with an 800 cfm capacity for measurement and a 15hp electric motor driving a positive displacement GMC 6-71 blower to move the air.

The selection of components for building your own flow bench can come from any number of places and the Internet and eBay are prime places to look for pieces that perhaps might work for you. There are even some very nice electronics packages available from some of the suppliers listed in the resources section that can help accomplish your goals and objectives without tanking your bank account.

You can follow the simple box approach or make something that is much more elaborate. You don't even need to have specific numbers to refer to as long as you have some relative pressure measurements so that you can see if you are gaining or losing airflow when you make changes.

# Chapter 5
# Flow Testing Tools, Measurements & Calculations

Airflow does some strange things. The little string flags allow flow visualization. Otherwise you would not know that some ports flow both ways at the same time. This stuff is not intuitive and testing must be done to find out the sometimes painful truths and surprises. Testing and thinking go hand in hand.

*No great discovery was ever made without a bold guess.* —Sir Isaac Newton

The study of aerodyanmics and the study of fluids are very closely linked. The discipline of fluid dynamics covers both subjects. Unfortunately, the study of fluids is like many things in life that are not intuitive. Instead of following intuition and guesswork, one must learn some of the rules of how airflow works in different circumstances and conditions.

## Basic Airflow Theory

Airflow is not an entity that is entirely on its own; however it does have some quirky likes and dislikes. It is perhaps easier to understand if we allow it to take on some human characteristics so that it is easier to learn its traits. Learning about airflow is certainly not intuitive and one needs to learn the particular rules that it lives by in order to get a better handle on dealing with its oddities.

Perhaps giving the airflow a personality and an entity of being might help to clarify some issues and make it easier to learn and remember. The following is a partial list of airflow characteristics:

• Airflow HATES to suddenly change directions.
• Airflow HATES to suddenly be expanded (as in a rapidly changing cross-section).
• Airflow HATES to suddenly be contracted (but not nearly as badly as above).
• Airflow HATES sharp corners at entries to runners or ports.

• Airflow LIKES very gradual directional changes in the flow path.
• Airflow LIKES very gradual expansions (as in a megaphone shape with small angles of divergence).
• Airflow LIKES very gradual contractions in cross-sectional changes.
• Airflow LIKES pipes used on exhaust ports.
• Airflow LIKES radii or liberally rounded corners at entries, like inlet guides.

It is not the intention of this section to present all the aspects of the study of fluids at the collegiate or graduate levels. However, it is the intention of this section to expose the reader to the many configurations of airflow measurement. There is much confusion about not only how to measure airflow through components, but even more on what to do with the collected data. There are more old wives' tales associated with airflow measurements than many other practices of similar contexts because the air can not be seen, so there is a fundamental distrust of data collection methods.

As an example, there are some rules about airflow that need to be learned and one of them is that just because two passages through a plate have the same area and are geometrically dissimilar, they certainly will not flow the same amount of air. It is because their $C_d$, or *coefficient of discharge*, are not the same. There are other fluids rules as well, but we will try and keep this more basic in the approach to better understand the world of fluid flow.

# Engine Airflow

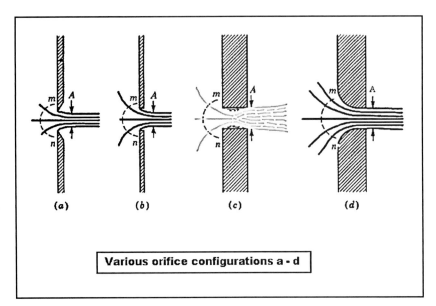

**Various orifice configurations a - d**

*Even though each of the orifices uses the same diameters, the coefficients of flow are not the same.*

This information is provided so that you can have some appreciation for how airflow functions. It is not necessary for you to know all about these things in order to become a better engine builder or a better racer or a better engineer. However, if you study more about the overall process, it will help you to become better at what you do.

## The Bernoulli Equation

Daniel Bernoulli was a Dutch-born mathematician who conducted a great deal of research on fluids, much of which was compiled in his book, *Hydrodyamica*, published in 1733. He was the first to ever measure blood pressure in the human body. His early work in fluids helped to establish a base for others to follow. The famous Bernoulli equation states that a fluid can exchange its kinetic energy for pressure. The relationships for fluid density, speed, and pressure are mathematically stated as:

$\frac{1}{2} \rho v^2 + p$ = constant
where $\rho$ = Greek letter rho (density), v = velocity, p = pressure

*Blood Pressure and Test Pressure*—You might not associate blood pressure with test pressure until you learn the process of measurement and how blood pressure measurements came to be. When you get tested for blood pressure, you can thank Daniel Bernoulli for the concept and equate it to establishing a test pressure measurement method that applies to airflow testing as well.

The blood pressure considered normal for adults contains two reference measurements. Systolic and diastolic refer to the high and low numbers of blood pressure measurement. The systolic is when the heart contracts and provides the highest pressure in the arterial system. The diastolic number is when the heart muscle relaxes and produces the low number in the arterial system. Normal blood pressure for adults is 120mm/80mm with each pressure referenced to the millimeters of mercury measured with a sphygmomanometer, sometimes called a pressure cuff. Think about the equivalents in pressure where:

120mm = 2.32 psi = 4.72 in.Hg = 64.19 in.$H_2O$; and
80mm = 1.55 psi = 3.15 in.Hg = 42.84 in.$H_2O$

All those equivalent pressures begin to put things in some better perspective. And at the same time your normal breathing rate is approximately less than 0.3 cubic feet per minute at approximately 3–4 in.$H_2O$ differential pressure ($\Delta P$). Maybe that information will help you win a trivia game sometime.

## Measuring Airflow

Measuring airflow through orifices, nozzles or a venturi can be done to exacting standards and provide adequate data on airflow capabilities of components or complete engines. Volume flow rate is often selected for ease of application. Calculations can be done to provide reference for a baseline of measurements. However, flow evaluations are not intuitive and one needs to learn the rules that apply in order to make a proper assessment. The most basic equation for airflow measurement is:

Q = AV
where Q = airflow in cubic feet per minute (cfm), A = area of orifice in square feet ($ft^2$), V = velocity in feet per minute (fpm)

It would be very convenient if things were truly that simple. The application of this equation can provide some useful information for general comparisons. Note, however, that there is not an allocation for the coefficient of discharge ($C_d$) for the orifice under evaluation. Thus, the data provided by the basic equation is lacking in specific detail.

In order to get better resolution with the calculation, we need to take a look at various $C_d$ values for different orifice shapes, as they are not all the same even if they are the same diameter. The simple equation of Q = AV becomes modified to be:

Q = A ($C_d$ K $P_1$)
where $C_d$ = coefficient of orifice, K = 4005 (this constant is for a sea level reference where the air

# FLOW TESTING TOOLS, MEASUREMENTS & CALCULATIONS

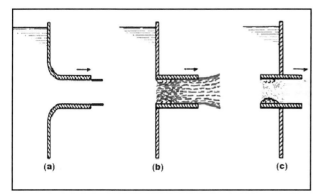

The coefficients of discharge for each of the profiles are not the same (a) 0.98 (b) 0.82 (c) 0.74 even though each example has the same inside diameter. Knowledge of these issues will help you to design better support systems for your race car or engine. Think about it. The edge thickness of an orifice makes a difference.

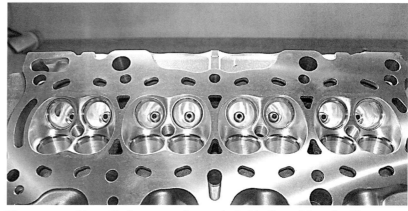

The same problems that you saw demonstrated in the entries of orifices based on shapes also happen in ports. Learning the reasons why airflow does things helps to make better port flow occur when you get a chance to reshape or evaluate a port. Photo courtesy Endyn.

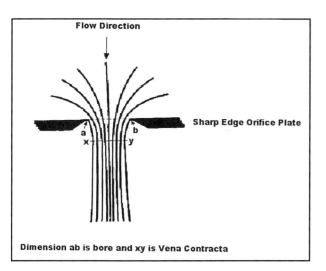

The diameter of the orifice, a-b, is not the same as the dimension of actual flow defined as x-y. The fluid flow phenomenon of the Vena Contracta creates the dimension x-y. The flow is reduced because of the Vena Contracta.

density is typically $0.075 lb/ft^3$), $P_1$ = pressure differential in inches of water

As an example, in the drawing shown on page 52, the coefficient for orifice (a) is 0.62 while (b) is 0.60 to 0.62, depending on diameter and thickness; (c) is 0.86, while (d) is 0.98. As stated earlier, the study of fluids and their measurement is not nearly as easy as it might seem.

Orifices have different characteristics that must be addressed in calculations or design of flow measurement equipment or experiments. Issues that must be part of the analysis are such things as the various coefficients of orifices. Coefficient of Velocity ($C_v$), Coefficient of Contraction ($C_c$), and the Coefficient of Discharge ($C_d$) are all related to the overall shapes of orifices. The $C_d = (C_v) \times (C_c)$. Coefficients for tubes and entry effect variations are shown in the photo above.

*Vena Contracta*—The coefficients vary and one of the reasons for the variations is the phenomenon of the restriction to airflow caused by the Vena Contracta, which is shown. The orifice shown is called a sharp-edge orifice.

The Vena Contracta would be exactly equal to the dimension of the orifice if the coefficient of discharge ($C_d$) was 1.00 or unity. However it is not unity and for sharp edge orifices and flat-plate orifices, this $C_d$ is approximately equal to 0.60 to 0.62 because of the flow phenomenon called the Vena Contracta and other issues related to the rules of flow through orifices.

*Reynolds Number*—A reference for flowing fluids as either smooth or turbulent can be described by using the Reynolds number. The Reynolds number was first used in 1883 and is named for Osborne Reynolds (1842–1912). The number is dimensionless and refers to the ratio of inertial and viscous forces in a fluid. The Reynolds number is important in identifying the characteristics of a flowing fluid. A Reynolds number of less than 2,000 indicates the fluid is in a laminar state. If the Reynolds number is between 2,000 and 4,000, the fluid flow is in a transition from laminar to turbulent. If the Reynolds number is over 4,000, the flow is considered to be completely turbulent. Laminar flow is called "streamlined" or "smooth" flow, and turbulent flow is called "rough" flow and is substantially more chaotic than laminar flow.

# Engine Airflow

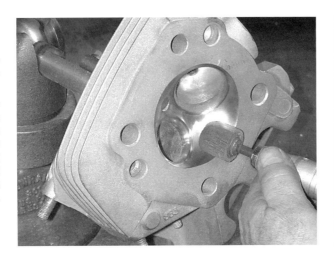

A Harley-Davidson head is getting the final treatment after the cylinder head has been reshaped. The work has revolved around getting an improvement in airflow without making the ports too large.

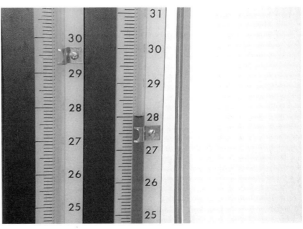

The test pressure on a flow bench can be displayed with either a vertical manometer as shown here or via other methods. The slight curvature that you can see in the photo is called the meniscus and the visual reading is taken at the bottom of the curvature. In the photo, the test pressure is on 28 in.H$_2$O, which is the standard in the industry.

Mathematically stated, the Reynolds number is:

$R_e = (\rho V D) \div \mu$
where $R_e$ = Reynolds number (dimensionless units), $\rho$ = Greek letter rho, density of fluid, V = mean velocity of the flowing fluid, D = diameter of pipe or orifice, $\mu$ = viscosity of fluid

The following equations can also be applied for airflow measurement, which is simply described as flow rate (Q) equal to the coefficient of discharge ($C_d$) times the cross-sectional area of the orifice times the velocity of flow. This applies readily to a sharp-edged orifice and is using metric units.

$Q = 0.1864 \times C_d \times d^2 \times \sqrt{h \times T_a \div pa}$
where Q = airflow in cubic meters per second, $C_d$ = coefficient of discharge, $d^2$ = diameter of orifice (mm) squared, h = head (mm H$_2$O), $T_a$ = temperature absolute (deg C + 273), pa = pressure absolute

As an example, let's apply the following equation to something such as a carburetor venturi:

$Q = [A^2 \div \sqrt{1 - (A_2 \div A_1)^2} \times \sqrt{(2gc \div \rho)(P_1 - P_2)}]$
where Q = ft$^3$/sec, $A_1$ = ft$^2$ area before venturi, $A_2$ = ft$^2$ area of venturi, gc = 32.17ft/lb-sec$^2$, $\rho$ = Greek letter rho (air density in lb/ft$^3$), $P_1$ = lb/ft$^2$, $P_2$ = pressure after venturi in lb/ft$^2$

A quick estimate of flow volume of air in cfm can be accomplished using the following equation:

$Q = 411 \times d^2 \times \sqrt{[(0.00102) \times (T_p)]}$
where Q = flow in cfm (cubic feet per minute), d = diameter of orifice in inches, $T_p$ = test pressure in in.H$_2$O

Many assumptions are made with this simple equation, one being that the fluid is air, another being that the $C_d$ is held constant at 0.60, so round orifices in thin materials are allowed for in this equation.

Another estimate of airflow through an orifice is the following:

$cfm = 13.29 \times d^2 \times \sqrt{P_t}$
where cfm = cubic feet per minute, d = diameter of orifice in inches, $P_t$ = test pressure in in.H$_2$O

Applying the following equation provides the calculation of local air density that can be used for simple airflow work. This equation will indicate how heavy a cubic foot of air weighs for the current atmospheric conditions. Sea level conditions of 29.92 in.Hg, 60°F provide an answer of 0.0763 lb/ft$^3$.

$\rho = 1.325 (P_b \div R)$
where $\rho$ = Greek letter rho (air density in lb/ft$^3$), $P_b$ = local barometric pressure in in.Hg, R = Rankine temperature (°F + 460)

Many other equations can be used to initially evaluate a system or components, but it is more common in industry to actually measure the components on an airflow bench that has been properly calibrated. It is most common to compare tested parts on a volume flow basis and at the same test pressures. Sometimes the requirement is to correct the volume flow data to mass flow data.

# Flow Testing Tools, Measurements & Calculations

Flow benches have a manometer such as this one to read the pressure differential across a known orifice inside the bench so you can have a cfm reference. This Saenz bench is rated at 600 cfm at 28 in.H$_2$O test pressure.

The inclined or flow manometer also forms a meniscus as this photo shows. The reading should be taken at the leading edge of the curvature when in operation. This one shows it is equal to 84.1% which would be multiplied times the flow range used for a cfm number.

*Test Pressure*—The test pressure is the primary reference to the differential pressure ($\Delta P$) that a device is tested. The test pressure is measured between ambient atmosphere and the other side of the test piece (cylinder head, manifold, throttle body, etc.). The normal reference for this measurement is in in.H$_2$O, and although the test pressure can be any number, it is very common to use 28 in.H$_2$O. You should always ask at what test pressure any flow numbers are given.

## Cfm As a Reference for Flow Testing

Cubic feet per minute (cfm) is the primary reference for the amount of flow, or *volume*, that a device produces at a given test pressure. There are other terms that can be used to reflect the mass flow or weight of air per unit time, but the most common term in use is the cfm reference.

There is an ongoing argument on what units the bench is really whispering to you and what gets put into magazines and books and what all that might mean to bench racing and power mongers everywhere. There are even some more various cfm references, but the ones already listed will be more than enough to get us into the fun side of trouble and controversy.

It is worthwhile to emphasize here that flow numbers without a reference test pressure are fundamentally meaningless. Always keep that in mind whenever you are looking at flow data or if you are collecting flow data. At all times you should have a flow reference at some given test pressure or the pressure at which the data was collected. At any time that you want to make a comparison of data you must have a reference for the test pressure. A 500 cfm data point sounds great until you realize that it was collected at 87 in.H$_2$O of test pressure. That equates to only 283.65 cfm at 28 in.H$_2$O of test pressure. Keep that thought in perspective or you will be lost following big numbers that don't mean that much when put in context of where you might normally accomplish your testing. There are several different types of cfm.

*Flow Bench Cfm*—This is typically the raw output of any flow bench that is normally referenced to cubic feet per minute (cfm) whether it is or not. You need to know where the numbers come from for this one in order to be trustworthy and confident of their origins. One might even refer to this flow bench output as units of flow if there are doubts about what that volume might be. If doing comparisons of components then units of flow vs the test pressure for those measurements is more than adequate for making meaningful comparisons. Simply put, three units of flow is greater than 2.5 units of flow. If this comparison

This Olds 455 head with mild bowl work was tested to flow 198 cfm at 28 in.H$_2$O of test pressure and at 0.600" valve lift. That is also 223 acfm and 13.38 lb/min under the test conditions. Calculate the scfm and see what you get. Read the text to learn about these meanings and how to do the calculations.

55

# Engine Airflow

This race car regularly produces ETs of about 7.3 to 7.4 sec at 177 mph in Denver (track elev 5,880 ft.). Assuming that it went to a track that was at a lower elevation like Las Vegas (2,020 ft.) and jetted correctly it would be quicker and faster with more dense air. Even though it is a bottle assisted car, local air density still has an effect.

was done at the same reference test pressure, then the gain is obvious.

Cfm refers to airflow either through the engine or that which a flow bench might measure directly. This measurement reference is also called *volume flow*. The cfm rating of components also must be accompanied with a pressure differential to establish a known reference. The pressure reference is often called test pressure and those units are typically given in inches of water.

*Acfm*—stands for *actual cubic feet per minute* and refers to the actual flow in cubic feet (airflow) that is for the local conditions at the time of a test. The acfm reference gets pretty difficult for some people to get their attention if they have never worked with airflow before. The acfm number from some benches might be a direct reading or it might come from a process of correction.

*Scfm*—stands for *standard cubic feet per minute* and refers to the airflow that is corrected to the conditions of a temperature of 60°F, and a barometric pressure of 29.92 in.Hg and dry air. Having stated that, there are many other references to scfm that are corrected to 68°F and 29.92 in.Hg, dry air. You need to know where the numbers come from and how the calculations are made. As you can hopefully understand, if the scfm reference is used, and the local conditions happened to be the same as the scfm conditions then there would be no correction needed. At that point, acfm and scfm would end up being the same numbers.

*Icfm*—stands for *inlet cubic feet per minute* and is typically used in reference to compressors for industrial uses, but is sometimes used erroneously. I won't expand or expound on this reference anymore other than to acknowledge that it does exist. In my opinion it has no use in discussing either engine airflow or flow through flow benches. So, now that you have been exposed to it, forget it.

Likewise, the term indicated cubic feet per minute would be the same as acfm. In order to avoid confusion, don't use the term icfm.

## Mass Flow

Mass flow is a reference that would be given in a number that refers to how much the airflow would be in weight per unit time. Here, I am using weight and mass as interchangeable for the sake of simple clarity. Essentially, the amount of air that a system uses or passes would have some mass and that mass per unit time would be expressed as pounds per minute (lb/min) or pounds per second (lb/sec) in engineering terms.

The reason that it is very common is because it is not cfm (volume flow) but the mass per unit time any engine will convert to usable power. However it is much easier to relate to volume flow (cfm) as a way to compare parts and components for internal combustion piston engines. The flow through those components will remain the same as far as their resistance to flow air. Remember until all the components are operating together on an assembled and running engine that those parts are not mass sensitive but the output of the engine is ultimately mass sensitive. So what that means is that an engine that is running in a race car at sea level (more density) will always produce more power than the same engine running in Denver, Colorado (less density at a mile high elevation). The size and capacity of the engine for airflow is still the same in both locations, but the power produced is sensitive to air density.

## Atmospheric Effects and Corrections

Atmospheric effects are of interest on some airflow benches but not on others. Do not fall for corrections that are not necessary.

Some airflow benches are designed to be ratio-based, volume flow devices. This holds true for many other manufacturers including the Jamison/Saenz flow benches as well. The design of the most common and the Jamison/Saenz flow benches is such that at all temperatures and pressures other than 60°F and 29.92 in.Hg (scfm conditions), the bench will provide data that is equal to the flow data that would be collected for scfm conditions. Only at the standard conditions of 60°F and 29.92 in.Hg will acfm and bench cfm be equal. So:

**Bench Flow = scfm = acfm at 60°F and 29.92 in.Hg**

thus no corrections are necessary for atmospheric conditions for bench flows as the output data from the flow bench is collected and displayed as if the

# Flow Testing Tools, Measurements & Calculations

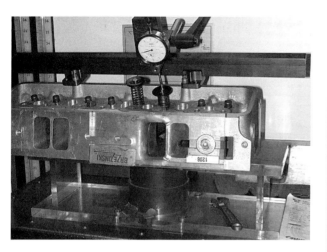

If you have measured the flow volume of something like this cylinder head and it flowed something like 335 cfm at 28 in.H₂O test pressure there is no real need to correct the data. However some folks do so and the text will explain how.

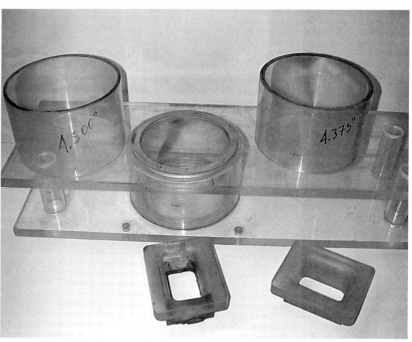

The bore simulator or bore adapter shown has replaceable bore dimensions and should closely approximate the bore the cylinder head will work with. The other devices are radius inlet guides which are also important in flow testing.

conditions were for those at 60°F and 29.92 in.Hg. The description holds true for reading the equipment manometers and is critical to know before attempting to accomplish any type of corrections. Is there any real need in correcting the bench readings to other conditions? Probably not, but some people will want to correct the data anyway. And it is necessary to understand where the numbers come from.

The following calculations can be applied so that the bench flow data can be corrected to a mass flow requirement or application.

Air Density (local) = $\rho_{local}$ = 1.325 [($P_{baro}$) ÷ ($T_{air}$ + 460)]
where $\rho_{local}$ = local density in lb/ft³, $P_{baro}$ = local barometric pressure in in.Hg, $T_{air}$ = local air temperature,°F

acfm = Q x $\sqrt{(29.92 \div P_{baro}) \times (T_{air} + 460 \div 520)}$
where acfm = actual cubic feet per minute, Q = flow bench flow in cfm, $P_{baro}$ = local barometric pressure in in.Hg, $T_{air}$ = temperature of air,°F

scfm = Q x $\sqrt{(P_{baro} \div 29.92) \times (520 \div T_{air} + 460)}$
where scfm = standard cubic feet per minute, Q = flow bench flow in cfm, $P_{baro}$ = local barometric pressure in in.Hg, $T_{air}$ = local air temperature in°F

Mass Flow = $Q_m$ = acfm x ($\rho_{local}$)
where mass flow = $Q_m$ = lb/min, Q = flow bench flow in cfm, $\rho_{local}$ = local density in lb/ft³, acfm = actual cubic feet per minute

## Useful Flow Bench Tools

The skillful use of tools will make the job of testing airflow much more fun, productive, and easier.

*Bore Simulator*—Supplies a simulated cylinder bore to mount the cylinder head to the flow bench. The bore dimension should be at least within +/- 0.030" of the bore size of the engine. The length of the bore simulator should be at least 4" to make it easier to use. The cylinder head should be positioned over the bore properly to duplicate the actual cylinder bore on the engine. There are advantages to having the cylinder bore made of clear material such as plastic so that you can see through it for some specific testing methods. The length of the simulated bore should be at least equal to the diameter or more, however there is rarely any gain for the length of the bore to be greater than about 4 to 4 1/2".

*Flow Balls*—These provide a method to probe a port and verify if flow is attached or separated at some point in the port. Flow balls are made by tack welding various diameters of ball bearings to a 1/16" diameter welding rod that is 12" long. Flow balls typically start with 1/8" diameter and go to 1/2" diameter in 1/16" increments. These tools are an easy way to find problem areas in the port. These simple tools are very effective for evaluating the short turn radius in a port or where a wall has a

# Engine Airflow

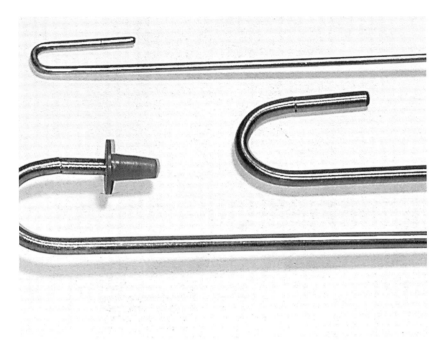

This device is called a Pitot tube in honor of Henri Pitot. A Pitot tube is a very useful tool for probing ports or checking flow activity in a flow stream. The Pitot tube will show you what you need to know about the elusive air in a port. Protective plastic cap keeps bugs out when not in use. See page 64 for details.

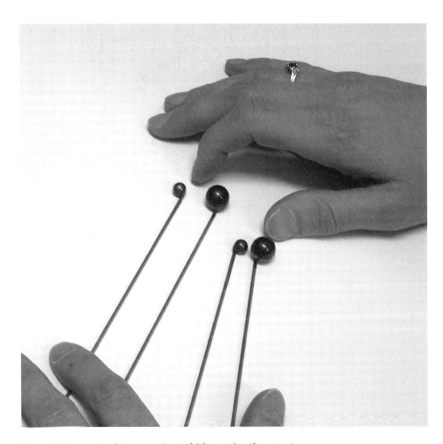

Flow balls are tools to use that aid in evaluating ports.

directional change. These devices help to reattach the flow of air if it has become detached. The flow balls also help to locate potential problem areas within a port or runner. For more on flow balls, see page 66.

*Port Molding Rubber*—This component provides an easy way to look at a port. The mold is made of silicone-based material that is poured into the port (with valve in place) and after it sets up, can be removed in one piece. The mold can be sliced and the cross-sectional profile drawn on graph paper to help measure the area at different locations in the port. The flexible rubber material allows you to change the shape or direction as part of the analysis process. With a port mold you can take a close look at changes inside the port, such as around head bolt column structures.

*Graph Paper*—Graph paper provides a simple and easy way to measure the cross-sectional area in a port. Cut-outs are trimmed to fit different places in the port and the squares can be counted for an accurate measure the areas. If you use the type of

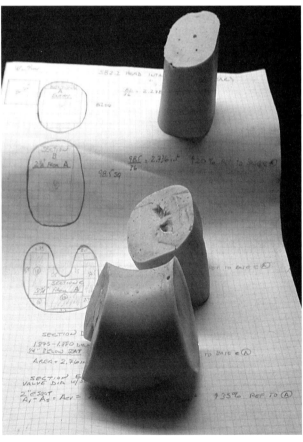

Port molding rubber isn't the most expensive support tool, but an easy way to take a closer look at a port and also generate cross-sectional data when used with graph paper. This is much easier than trying to transfer measurements by other methods.

## Flow Testing Tools, Measurements & Calculations

Use a radius inlet guide for the airflow to smoothly turn the corner into the port. Otherwise an "edge effect" takes place and decreases the airflow number considerably. Sometimes the loss can be anywhere from 6% to 10%. Do not flow test ports without a radius inlet guide.

This somewhat proves that some things that you were exposed to as a child are still worth remembering. This type of modeling clay is ideal for shaping and forming ports when testing on the flow bench.

graph paper that has 10 squares to the inch (100 squares to the square inch), you can get very accurate measurements by simply counting the squares that are within the outline of whatever piece or section of the port molding rubber that is under investigation. Careful layout and counting allows very good resolution of data without much complexity.

*Poster Board*—Poster board can be used to make patterns to help the developer reproduce the same port shape and size. The poster board can then be used to trace out patterns in aluminum or plastic to duplicate an established shape in other ports and cylinder heads of the same type.

*Radius Inlet Guide*—Provides a smooth approach to the port or device that is being tested and is intended to decrease the "edge effect" at the port flange. The radius used should be as large as possible, at least 1/2"–3/4". The thickness of the inlet guide should be at least 50% of the height of the port. The size outside the port cross-section should also be at least 50% of the height of the port so that there is a smooth approach in all directions. It is not uncommon that the inlet guide will improve the flow by 6% to 10%.

A liberally sized radius inlet guide is a very effective tool. Most people make them too small and use too small of a radius. A rigid radius inlet guide should be used whenever possible instead of using modeling clay. The clay is quick, but is difficult to duplicate the same liberal radius from port to port testing.

*Exhaust Pipe*—All testing of the exhaust side of the cylinder head should use a short section of exhaust pipe that is at least the diameter of the port.

The appropriate length is about 10" to 12" long.

*Modeling Clay*—Clay can also be used to change shape or to fill in a port in some space. It is not uncommon to find that filling in the floor of a port might not hurt the flow at all because not a great deal of air runs along there. At least you can check it out without having to weld and cut and then see what the effects were. Clay is very easy to use for smoothing and blending in transition areas of the port or runner. More often than not the clay should be reached for long before consideration is given to grabbing a grinder.

*Flow Wand*—Provides an indication of activity and direction of flow inside a port or device under test. The magic wand is made from a 1/16" diameter welding rod that is about 12" to 14" long. The welding rod has a small round ball on the end (from placing the rod in a molten pool of metal) and a piece of Dacron or nylon type kite string is glued to itself to form a little flag about 3/8" to 1/2" long. There is no need to use longer lengths. The little ball on the end looks like the top of a flagpole. The flow wand (string flag) provides a pretty good flow visualization indicator. Because

This little flow indicator string flag (flow wand) is a handy tool to show you where the activity in the port is and also which direction the airflow is going. They are commercially available or you can make your own. Flow visualization tools such as this are easy to use.

# Engine Airflow

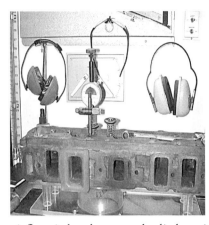  

There are many types of valve opening devices for use on flow benches. The first photo is a direct acting indicator/opener from PMS and the middle photo is of a nice piece from Cal Spec. The last photo is of a digital indicator/opener that can be used to open the valve automatically.

airflow is hard to see, the little string flag allows you to note what direction and activity is produced by the airflow where it is used. No numbers, just motion and direction. For more on using a flow wand, see the sidebar on page 70.

*Smoke Candles and Smoke Generators*—Provide flow visualization techniques that can help the operator "see" where the airflow goes and how it is affected by changes in directions and restrictions or expansions. The problem is often what to do with the smoke as it exits the flow bench. It will fill up a small room rather quickly. If you are flow testing at a volume flow of about 250 cfm and the room is only 1,000 cubic feet, it takes only four minutes for the air to fully exchange including the smoke. So think carefully about the process of how to deal with the excess smoke before you cause a scare for everybody else in the place.

*Wet Flow Adapters*—These provide another reference of flow visualization that can prove to be valuable in evaluating what is happening in the combustion chamber and helps to sort out problems in that area. It is best using this process when at an air/liquid ratio that represents the normal air/fuel ratio that an engine uses. This process has helped to solve some problems that otherwise would have gone unnoticed. The legendary Joe Mondello and his partner Lloyd Creek should be appropriately credited with making the first wet flow attachments for flow testing commercially available.

*Calculator*—Should be always present when the flow tester is on the telephone comparing numbers with another tester. A calculator provides an instant indication of hype vs. truth as known numbers are compared to claims. The numbers game can be a challenge, but used in the correct fashion, turns out to be a sort of truth serum. Almost any inexpensive calculator will work, but try and use one that has a square root ($\sqrt{}$) key. It is also more convenient to use one that has a 1/x key for quick reference for reciprocal values. When I first started working with these type equations and calculations, the common calculator was a slide rule and electrically powered calculators were laboratory-only items. The hand-held calculators today are so much easier to use than a slide rule and they even have memory keys.

*Valve Opening Attachments and Dial Indicators*—Some method is necessary to position a valve so that the operator can reference its lift point. There are many ways to do this task. Some of the attachments are available from Precision Measurement Supply. Bill Jones in Utah has been providing flow bench accessories for many years and has a very good selection of inexpensive tools for flow bench work. The opening device needs to be easy to use and repeatable. Most valve opening devices also use a dial indicator for a quick and accurate reference. Be aware that many dial indicators are either direct read-out or might be using a ratio. Remember the part about needing to know where the numbers come from?

Wet flow adapters are another way to "see" what is happening with the flow in the area of the combustion chamber and around the valve seats. Liquid that is colored with dye helps to point out problem areas. This kind of test gear can be fit to any flow bench. This type of rig gives indications, not specific answers.

# Flow Testing Tools, Measurements & Calculations

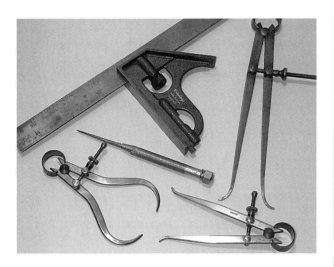

Layout tools are necessary items around airflow projects. Scribing and measuring is something that you will get used to. Missing from the photo is a felt marking pen and Dykem layout fluid.

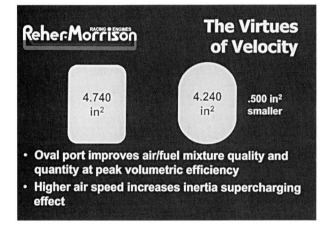

This is a comparison of different port cross-sections and shapes that produced good results in BBC engines. Big power and big airflow does not mean the ports have to be big. Photo courtesy Darin Morgan.

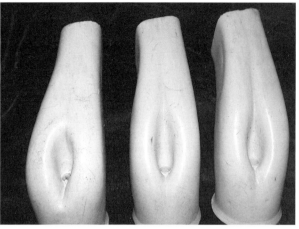

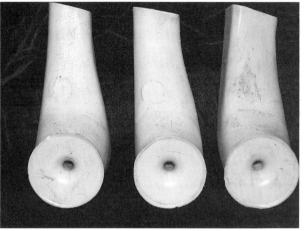

These port molds are from various configurations for an SB2.2 Chevy engine several years ago. Some of these heads flow better than 420 cfm at 28 in.H$_2$O. The ports move a great deal of air while maintaining a small cross-section, which results in high velocity.

## Measuring Cross-Sectional Areas in Ports

It is a common mistake that many engine builders and cylinder head modifiers make when they simply multiply the height of the port times its width. Use of that calculation for area without considering the effect that either an irregular shape or the radii in the corners have on the final answer generates errors in engine program computer simulations, calculations, and evaluations.

The majority of typical ports are somewhat rectangular with various radii used in the corners. This rectangular port configuration is the most common for intake ports while exhaust ports may use the rectangular, square or variations in most cylinder head applications. Some ports are shaped in a somewhat oval configuration. Some port shapes use the generally rectangular configuration, but have a different radius in each corner and are irregular in shape. Some port configurations might also look a great deal like church windows or other complex shapes. Regardless of the shape of the intake or exhaust ports, it is possible to measure them and do so with close accuracy. Learn about the cross-sectional area so that you can make good decisions on port velocity and other factors using the methods that follow.

Accurate area measurements of ports, runners, and manifolds are normally very difficult unless one has the proper instrumentation. However, the following methods will make the process easier.

The description of each method of measuring and calculating the port area will also present the conversion to square feet so that it is easy to use with flow bench data, which is normally in cfm (cubic feet per minute).

*Port Velocity*—Port velocity might seem more like a buzzword to the majority of engine builders and cylinder head modifiers. The very few that understand and properly apply the knowledge to attain correct and complimentary port velocities

# Engine Airflow

Somewhere down in there is a cylinder head on a flow bench. Airflow numbers need to be gained with all the parts attached as in this case where some fuel injection parts are attached in the flow path.

pick up free power and build engines with superior characteristics (power range = the rpm spread from peak torque to peak power).

What is the velocity and how high is too much? A good maximum port velocity target in a running engine should be no greater than 55% to 57% of the speed of local sound. This can also be referred to as 0.55 to 0.57M. The Mach number references the speed of sound and is named for Ernst Mach (1838–1916). Mach was a physicist, philosopher and early pioneer in studying the speed of sound. The speed of sound is 1087 ft/sec at 32°F with an increase of 1.1 ft/sec/°F. The calculation for the speed of sound (Vsound) in ft/sec is approximately:

**$V_{sound} \approx 1052 + (1.1 \times T)$**
**where 1052 = an adjustment constant and T = °F.**

There is also a theory that targeted airflow velocity *in the port* should be about 200 to 250 ft/sec (in the running engine), which would only be about Mach 0.17 to Mach 0.22. After reading some of the information in this book, you can decide which approach best satisfies your development and testing goals.

As an estimate, the temperature at inlet valve closing for a naturally aspirated engine in the port is approximately 100°F. So, using 100°F would produce a speed of sound in the port of 1162 ft/sec and 55% of that speed is 639 ft/sec. Now that particular temperature is not a perfect number for scientific studies, but will get you close enough to apply the information in engine planning and calculations. It turns out that even in forced induction (supercharged or turbocharged) engines, the 100°F reference will also work if they have an adequate intercooler or aftercooler. In supercharged applications, if you know the temperature in the manifold (just prior to the intake valve) you can use that temperature reference by applying the simple formula given previously.

At this point, note that when considering the airflow of an engine, there is no difference in manifolding, connecting runners, or ports in the cylinder heads, as they are all an extension of the same flow path. Also note that the various differences in cross-sections (areas) allow different local velocities. If we look carefully at the geometry of the airflow path within the engine, it should become very apparent that the smallest area will yield the highest local velocity. Evaluating the airflow path of the engine and averaging the lowest velocity with the highest velocity will allow a consideration of average velocity, which truly sets the characteristics of the engine, but it is the *minimum* cross-section that establishes the maximum velocity.

Another calculation of port velocity without having anything placed in the flow path is from the equation:

**$P_{vel} = (P_s \div 60) \times (B^2 \div A_p)$**
**where $P_{vel}$ = port velocity in feet per second, $P_s$ = piston speed in feet per minute, B = bore diameter in inches, squared, $A_p$ = area of port in square inches.**

So, when you hear someone talking about how the speed in the intake or exhaust system was sonic (at least Mach 1) consider that is very, very fast. If 55% of the speed of local sound (at 100°F) is 639 ft/sec and 60 mph = 88 ft/sec, then that is 639 ÷ 88 = 7.26 x 60 = 435.6 mph. Pretty difficult at best. Sonic would be at a local speed of 792.27 mph in the port! A speed that is not very likely at all. However the exhaust port might have extremely brief sonic flow at very low valve lift during the early occurrence of exhaust blowdown when the residual pressure in the cylinder is still high compared to the ambient atmosphere. The speed of sound at the exhaust port (about 1,200°F) would be approximately 2,370+ feet per second.

The area of a square or a rectangle is width (W) times the height (H). In the case of a square, both sides are the same, but using the H x W approach will work. The real problem occurs in the corners where there are radii used to form the port of the manifold, runner, or duct in the flow path.

Because the calculation for the area of a circle is well known, then it can be applied to calculate the equivalent area of either a rectangular or an irregularly shaped port cross-section. Even egg-shaped or church window–shaped ports can be accurately measured for cross-sectional area with this method.

The measurement of the area would be very easy

# Flow Testing Tools, Measurements & Calculations

by using a planimeter, but they are very difficult to find and are very expensive. It is even a lot easier if you had access to a CMM (coordinate measuring machine) but that is very expensive and time consuming.

***Rectangular Port Cross-Sections (Method 1)***—
It is always easier to evaluate and analyze a problem if you make a sketch. It also helps make sure all the details are considered. The equation to calculate the area of a port with the following dimensions H, W, $R_{cnr}$ is:

$(H) \times (W) - A_{cnr} = A_{prt}$
where $A_{cnr}$ = area of corners, $A_{prt}$ = area of port (in$^2$), H = height of port, W = width of port. $R_{cnr} \times 2 = D_{cnr}$. $D_{cnr}^2 \times 0.7854$ = area of circle for the corner radii. The diameter of the circle scribed using the radius of the corners can be placed in a square of the same dimension; the area of the circle subtracted from the area of the square represents the area of the corners ($A_{cnr}$).

Example: Where (H = 2.150", W = 1.060", $R_{cnr}$ = ½ inch). Where the radius of all four corners is the same:

Step 1: H x W = 2.279 in$^2$
Step 2: $R_{cnr}$ x 2 = $D_{cnr}$ = ½ x 2 = 1
Step 3: $D_{cnr}$ = 1, so $D_{cnr}^2$ = 1 x 0.7854 = 0.7854 in$^2$
Step 4: 1 – 0.7854 = 0.2146 in$^2$ = $A_{cnr}$
Step 5: H x W – $A_{cnr}$ = $A_{prt}$ in$^2$ = 2.279 – 0.2146 = 2.064 in$^2$
Step 6: $A_{prt}$ ÷ 144 = 2.064 ÷ 144 = 0.0143 ft$^2$

***Port Cross-Sections (Method 2)***—The periphery of the cross-section can be measured with either a string or stiff wire. By carefully forming the string or stiff wire into all the radii of the corners (even if none are the same) an accurate measurement of that section can be done. Measuring the string or stiff wire provides a length ($C_{prt}$).

$C_{prt} = \pi D$,
where $C_{prt}$ = circumference around the port, $\pi$ = 3.1416 and D = an equivalent diameter. When D is found, $D^2 \times 0.7854 = A_{prt}$ in$^2$. $A_{prt}$ ft$^2$ = $A_{prt}$ ÷ 144.

Example: A port with four different, but unmeasured, radii in the corners is measured with a stiff wire and the resultant length is measured to be 6.3". What is the area of the port at that location?

$C_{prt}$ = 6.3, so 6.3 ÷ π = 2.005
$D^2 \times 0.7854 = (2.005)^2 \times 0.7854 = 3.16$ in$^2$ = $A_{prt}$
$A_{prt}$ ÷ 144 = Aprt ft$^2$ = 3.16 ÷ 144 = 0.022 ft$^2$

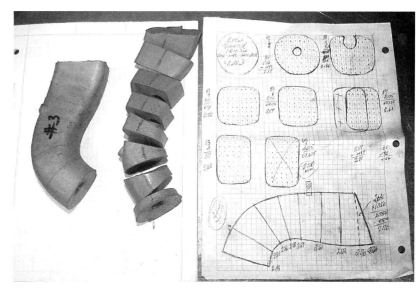

This method of port cross-sectional measurements is very accurate and gets the details of various places in the port identified. This is a great method for evaluating the port at different locations. Photo courtesy 10 Litre Performance.

***Port Cross-Sections (Method 3)***—This method is sometimes called the "paper doll" method. Although it sounds simplistic, it is a very accurate method. It uses graph paper cut-outs to measure the different sections. The preferred graph paper to use is the type that has 10 squares per inch (thus 100 squares per square inch). The graph paper is cut out so that the graph paper is an image of the port cross-section to be evaluated and the squares can be counted. Those squares on the periphery of the cut-out that are less than one square are estimated for a value of 1/4, 1/2, or 3/4 of each square for greater accuracy. The total squares are counted and then divide by 100 for the $A_{prt}$ in square inches. The $A_{prt}$ ÷ 144 = area in square feet.

Example: Graph paper using 100 squares per square inch is used to make a cut-out of the port cross-section 2" in from the flange surface. The number of little squares counted is 237 (allowing for some around the periphery that were not full squares). What is the area of the port at the location described?

237 ÷ 100 = 2.37 in$^2$ = $A_{prt}$ in$^2$
$A_{prt}$ ÷ 144 = Aprt ft$^2$ = 2.37 ÷ 144 = 0.0165 ft$^2$

***Port Cross-Sections (Method 4)***—This method depends upon molds being taken of the port. Various schemes of taking mold impressions have been done for ports, but a favorable process is to use mold maker's rubber (available from several sources). The molds can be sliced into sections at several reference points and then the mold section or slice can be placed on graph paper as in Method 3 above and the squares counted and the area is calculated in the same manner.

*Port Cross-Sections (Method 5)*—The port can be measured by placing the part (cylinder head, manifold, runner, etc.) on a CMM (coordinate measurement machine) unit and the resulting three dimensional "digitizing" of the part can be used to calculate the areas and even volumes and lengths. CMM time is very expensive and measurements from this approach would be cost prohibitive in most circumstances.

The use of CNC milling machines for port shaping allows multiple calculations to be done and supplied with the finished parts. However, this method is not common for most customers of CNC porting shops because if the shop supplied the data with the cylinder head or manifold, they might be giving away part of their measurement and machining technology which most might consider as proprietary for a given port design. Quite often the secrets of CNC machining relies on the tooling as much as the programming skill of the operation.

## Flow Bench Data and Port Area Measurements Provide Local Velocity

As a matter of convenience for this evaluation, many airflow benches provide flow data directly in cubic feet of air per minute (cfm). Since we are interested in velocity, it is imperative to know the area of the port. If the port area was expressed in square feet ($ft^2$), then dividing the flow value in cfm by the area in $ft^2$ (square feet) of the port, the result is FPM (feet per minute). Dividing FPM by 60 yields FPS (feet per second). The specific advantage of this method is that it is not invasive (puts nothing in the flow path) to the port and is a quick reference. See also the effective flow area evaluation on page 67.

## Port Velocity Using Flow Bench Data and Port Area

Flow bench data = At 28 in.$H_2O$ test pressure a port flows 350 cfm, port area in square feet = 0.022 $ft^2$ (as described in Method 2 on page 63). Local velocity at point of area measurement = 350 ÷ 0.022 = 15,909 FPM ÷ 60 = 265.15 ft/sec. Remember that 88 ft/sec is equal to 60 mph? 265.15 ÷ 88 = 3.01 x 60 = 180.8 mph! Right there on the flow bench. It sort of puts things in some perspective doesn't it?

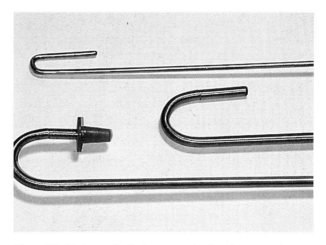

Three Pitot tube ends that you see in the photo are the ends that are placed into an airflow stream. The larger ones are 0.125" in diameter and the smallest one is 0.065" diameter. Look closely and you can see the static ports, which are small holes around the outside of the tubes. They each have a small hole in the end that is the impact port that must face the flow. Measuring the pressure differential between the static and the impact ports gives the total pressure.

## Measurement of Local Velocity Using a Pitot Tube

This type of measurement can be done but the nature of the Pitot tube can cause problems. The Pitot tube is sensitive to yaw (angle of airstream other than parallel with Pitot tube) relative to the standard direction of the developed flow path in a port. It is also very difficult to measure the local velocity with a Pitot tube in the vicinity of the short side radius (where the flow path turns from the main stream to the area below the valve). This measurement can be done directly if using a flow computer with a matching Pitot tube. The indicated local velocity can also be read with a vertical manometer (using the instructions supplied with the Pitot tube) or by use of the flow computer instrumentation set to read flow velocity directly. The only type of Pitot tube that is less sensitive to the yaw angles that occur in the port or runner is the Kiel Probe. Kiel Probes are very expensive little devices but allow much more yaw angles of up to 8 degrees (angle from parallel to the airflow path).

## Flow Testing with Pitot Tubes

The design of the Pitot tube is essentially a tube within a tube. The impact port (stagnation port or point) is the central tube. Surrounding the inside tube is the portion that provides the housing for static ports which are tiny holes drilled (equally spaced) around the outside (circumference) of the tube housing. This configuration is common for all

# Flow Testing Tools, Measurements & Calculations

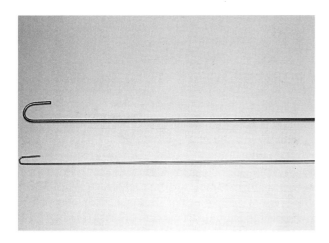
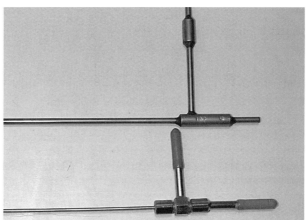

**The Pitot tube has a connection point for two sides of a manometer. The end of the Pitot tube connects to the bottom of a well type manometer and the side connects to the top of the manometer. Simple calculations will give you air speed.**

types of Pitot static tubes.

Even though any tube will give signals to a manometer, the Pitot tube must be connected in the manner described in order to make any sense of the numbers generated on the manometer or pressure transducer.

***Tubing Connections to a Manometer***—Connecting to a vertical manometer is easily done. Tubing connections are provided in the construction of the Pitot tube. Properly sized tubing can be used to connect the Pitot tube. The connections to the manometer must be done in the following order:

The straight section (impact port is a static + dynamic reading) must connect to the bottom of the vertical manometer. The 90-degree section (static ports) of the Pitot tube connects to the top of the manometer. When properly connected with the tubing supplied, the Pitot tube will produce a reading on the vertical manometer that is usable in calculations for local velocity.

You can easily make a Pitot tube to use on the exhaust system. It can be a straight section of tubing or it can be bent at 90 degrees. The intake Pitot tube is more problematical in that it has to be a tight bend of 180 degrees. It is much more difficult to use effectively in the intake ports because of its size and in that it gets in its own way (airflow shadow). Your head can also have an effect on the numbers if you get too close to the port runner or carburetor or throttle body on the bench.

***Using the Pitot Tube and Manometer with Calculations***—The Pitot tube readings can be used to calculate local velocities in applications other than ports or ducts. The methodology of calculation is air velocity ($V_{air}$) is a function of the square root of the velocity pressure ($P_v$) divided by the local atmospheric density ($D_{air}$) multiplied times a constant (1096.2) so that the air velocity is expressed in FPM (feet per minute). Corrections are necessary for conditions where data is taken at other than 70°F. The calculations should yield accuracy of local velocity within +/- 5%, assuming no problems with excessive yaw of airflow relative to the impact port centerline.

$$V_{air} = 1096.2 \sqrt{(P_v \div D_{air})}$$

**where $V_{air}$ = velocity of air in fpm, $P_v$ = velocity pressure in in.$H_2O$, $D_{air}$ = density of local air in lb/ft$^3$**

One can simply divide the fpm number by 60 and the result would yield feet per second (ft/sec),

**This student is probing an intake port with a Pitot tube to map the port for velocity variations. The Flow Lab at Ranken Technical College he is working in is a great place to study airflow. He is working with an automatic flow bench rated at 1,000 cfm at 28 in.$H_2O$ test pressure. Photo courtesy Ranken Technical College.**

which is a common reference for port velocities. You can even use this calculation for using a Pitot tube for a speedometer (as in aircraft use) and you would have an IAS or indicated air speed and corrections would be done for variables that get the user to a TAS or true air speed answer.

$D_{air} = 1.325 (P_{baro} \div T)$
where $D_{air}$ = density of local air in $lb/ft^3$, $P_{baro}$ = local barometric pressure in in.Hg, T = absolute local temperature (°F + 460)

## Using Flow Balls to Identify Separation Points in a Port

Probing a port with various sized flow balls such as those made by Thorpe Engine Development will help locate the local flow separation in a port. Some cylinder head developers prefer to use the Pitot tube or other pressure measurements to accomplish this goal, but flow balls are actually much easier to use. The ball is placed in various points in the port being tested on the flow bench and when the flow has separated the placement of the ball will help to reattach the flow and the test pressure will go down (thus showing a flow gain). Then you need to evaluate how to make the port flow to stay attached with either small obstructions or reshaping. Attached flow is always more preferable than separated flow.

*Flow Balls Testing Procedure*—Flow balls are a tool that upset the airflow, positively or negatively. We use *both* the positive and negative effects on airflow to learn and help us visualize what the air (which we can't see) is doing. From that, we can respond accordingly.

To begin, let's graph lift vs. flow. Start at 0" lift and don't forget to record leakage. Now we have a baseline. Leave the valve at full lift and we'll initially start using our flow balls here.

When you insert a flow ball into a port, hold the rod parallel to the wall because you don't want the rod upsetting the flow. Watch for changes in flow and the magnitude of the change. Did it change a lot? That area is very "active." Changes made there will affect airflow, positive or negative. Did it change a little? Changes made there probably won't yield much return. Did it not change at all? That indicates a "dead" area, a good place for fuel drop out, so maybe it should be filled. Try filling it with clay and note any change.

Think of it like this: On a windy fall day, where are the leaves? In the air! What happens when the wind stops? The leaves drop to the ground. Where do the leaves collect around your house? Where the wind doesn't blow! So fuel collects at those dead areas. It might be on the wall behind the pushrod choke area, or on the floor or in a corner. You've got to activate those dead areas. How? I can't tell you; every port is different. You've got to be smarter than the air. Try to "set up" the air to go where you want it to go, upstream of that area/point. As you're moving your ball around you will feel the turbulence on the rod, you will sometimes hear a pitch change, and you will see a flow loss or gain. That is indicating a spot that needs some work. We use our ears to hear, hands to feel, and eyes to see the changes on the manometer, then our minds to visualize the airflow. It is a dynamic testing procedure. We are testing while air is flowing.

An ideal port is one where in any position, at any point, the flow ball will hurt the flow. This indicates that all surfaces are working. That is what we are aiming for. What size ball should hurt the flow? When the smallest ball hurts the flow the better our port is performing.

At a fixed distance inside the port, move your flow ball all around the port. Quite likely you'll see one side or an area that doesn't flow as much as the other. The ball will cause a reduction of flow on one side and no change at another point. The reduced flow results from the ball blocking, impeding, or killing the flow, while no change of flow indicates the ball is located in a stagnant or dead area.

Hold your flow ball very lightly between your fingers; a 5/16" or 3/8" ball works good. Now enter the port and feel where the air is taking the ball. Probe the port all around but guide the ball very gently, let the air take the ball. Where is the air taking it? It is being shoved this way and that. Depending on the ball size, it will vibrate back and forth at certain spots. In certain areas the ball just does not want to be there! Can you visualize the air stream as it carries the ball? It is somewhat like holding a kite and watching where the air stream takes it. It will give you a good overall picture of and feel, through the rod, of the flow dynamics of the port. In this example you are using the ball as a kite.

Flow balls are a perfect tool to check corner radii. Start with your biggest ball at the mouth of the port and push it in until you get a change in flow. Then go to the next smaller, etc. Can you visually see in your mind an envelope area? Now do this to the other three corners.

Next check the floor; place your flow ball, again starting from largest to smallest, dead center on the floor. Slide your flow ball in until you see a change. Now do the sides and the roof. You are "port mapping" this port in your mind. Can you see what is happening in all four quadrants? What happens as you move towards the valve? This, combined with flying your flow ball kite, should give you a pretty good indication of where the air is in the port and how fast it is going.

While you were mapping the port, you should have heard, with your ears, a change of pitch. That is telling you something! Now here is where we have to figure out what the ball is doing. Visualize a river with a rock in it. What is happening at the front/sides/back of the flow ball, or rock? Pressure is building at the front and flow is separating at the trailing end. The flow separation creates vortices where the flow at the trailing end of the rock is turbulent. If we put another rock behind our original rock that would help straighten the flow. This is where it gets tricky. Or, if we put a larger rock in front of our original rock, the smaller rock behind our larger one will help "straighten" the flow. Which is it? Try the different size balls to determine which it could be. Does your river of air sound smooth or are there lots of rocks and waterfalls in it?

What about when the port sounds really good up to a certain lift point then starts sounding harsh up to maximum lift? Don't try to fix it at maximum lift; reduce your valve lift to where it just starts to sound harsh. That harsh sound is when the air is separating from the wall. Fixing it at this point could cure it at maximum lift. Start there and work your way to maximum lift. It should sound smooth and consistent.

If your test pressure is fluctuating all over the place it's usually caused by an unstable situation as the air approaches the valve seat. The noise will vary with the test pressure; as the test pressure goes up, the noise will go up, and vise versa. The air is separating, then reattaching. As the test pressure goes up, the flow goes down (inverse relationship). Try probing with different size balls. What area or position in the port did the ball cause the flow to stabilize?

If you put a ball in just before the short turn and you pick up flow it might be telling you; the vortices created behind the ball is reattaching the flow. Did the sound transition from harsh to smooth? It should have. Or there might be something upstream that the ball is straightening out. The "trip" off the back of the ball might be causing an attachment, or the front of the ball might be acting like a small rock behind the larger rock. Probe the port with the different size balls in that area moving in and out, up and down. This is not limited to the short turn. Flow might be separating around a corner at the push rod. Or it might be that your air speed is too high. Check with a Pitot tube; 380–400 feet per second is too fast (38 in.$H_2O$). If your speed is too high, slow the air down by increasing cross-sectional area. Where? I can't tell you, but don't do it in a "dead" area. If your speed is okay then the port shape could be the problem, or it might be in the combustion chamber.

If you have a problem that the flow ball "fixed" try going upstream with different size balls. Your port should sound smooth and consistent. Always work as far upstream of your problem as possible. Sometimes that will cure two problems downstream in the airflow path.

Now let's look at the flow vs. lift graph. Are there any dips or irregularities? Set the valve at that lift and probe the port with different size flow balls. Probe to look for areas of sensitivity. Watch the position of the ball where it increases the flow and where it has no effect on the flow. Analyze it. Is it the front or the back of the ball that is turning/tricking the air? This would be a good time to use your "flags." Use the 90-degree rod and drag the ball end on the port wall. While working your way towards the valve, zigzag across the floor, then sides, and roof watching the string. Is it in line with the port? Or at a certain point does it change direction? Does the string stay attached to the wall or does it lift off? If it lifts off, where does it point? What size ball can you put in at that point that will hurt the airflow? Try and pretend you're an air molecule, and you hate to turn corners.

I know it takes more time, but try to record the effect of each test. Yes, you wind up with LOTS of data. But all that data can be useful! It's telling you a story, and you can determine the ending of that story! So when you can't figure out what the next page should read...take a break! Then start fresh and new tomorrow.

## Method for Calculation of the Effective Flow Area of Ports and Orifices

Based on a mathematical equation, this calculation depends upon very accurate data from a flow bench, and appropriate and complex calculations can be applied to calculate and establish an effective flow area (EFA, $A_e$) for the port or orifice. When the $A_e$ is compared to the physical measurement of the port area in square inches, a correlation can be done to rate the port. The port area is typically measured on a CMM (coordinate measurement machine) with this method, although other methods of cross-sectional measurement can be done. This is another way to rate ports with a Coefficient of Flow ($C_f$). This methodology is normally used only in the realm of engine designers and engineering personnel

# Engine Airflow

Tiny holes all around surface

This is not an alien spacecraft although it looks like one. The thousands of tiny holes in this gas turbine part can be measured quite accurately on a flow bench if the correct process is applied. The tiny holes flow air to cool the part so it won't warp inside the running engine. Flow benches can be used for lots of strange projects.

assigned to process data for analysis using CFD (computational fluid dynamics). However, more and more gearheads are getting involved in the higher level analysis of airflow pathways. Many shops that have CNC (computer numerical controlled) equipment also have the capability to very accurately measure the cross-section at various points in the ports of cylinder heads, manifolds, and throttle bodies with ease.

The ratio of the geometric flow area to the effective flow area is the $C_d$ of the port or orifice. Mathematically stated:

$$A_e \div A_g = C_d$$

where $A_e$ = effective flow area, $A_g$ = geometric flow area, $C_d$ = coefficient of discharge.

## Fleigner's Equation and Airflow Applications

The initial and earlier work of Saint Venant was verified by Fleigner in his experiments and work done in 1874 to 1877. The time was very important in the development of steam engines, spark ignition engines and airflow. By the time the 1900s rolled around, Fleigner's equation was commonly used by anyone studying both thermodynamics and fluids and steam engines. In modern times, the Fleigner equation is commonly used to evaluate flow from one volume to another via a small passage or "leakage" paths. Fleigner's equation can readily be applied to many compressible flow conditions and even to immersed bodies. The basic formulation of the initial Fleigner's equation is:

$$Q_m (C_p T)^{1/2} \div A (p_s + \rho V^2)$$

where $Q_m$ = mass flow rate, $C_p$ = specific heat at constant pressure, T = temperature, A = area, $p_s$ = local static pressure or pressure drop, $\rho$ = mass density, V = velocity or flow speed. The form listed above can be a dimensionless number.

Next is how to apply the equation to flow benches and make a comparison of parts (such as various gas turbine or other components). In my experience, these methods are acceptable to most of the contacts I have in the aerospace and industrial gas turbine industries, including Allison, Solar, Garret AirResearch, Allied Signal Garrett, Lycoming, Pratt and Whitney (Canada), and others. The methodology is simple and very cost effective. There is a consideration for the *geometric flow area*, which is the area calculated from the areas of the passages or channels or orifices, and there is the consideration for the *effective flow area*, which is the area indicated by actual flow testing. The simple coefficient of discharge ($C_d$) of the device or part is equal to the effective flow area divided by the calculated geometric flow area.

Some fundamental equations can be used with flow benches to calculate the Effective Flow Area (EFA) using Fleigner's equations. The methods listed produce data in English Engineering units and the presentation of the formulae is following the use of AOS (algebraic operating system) so that they can be used on any simple scientific calculator. Most flow benches have flow output data in volume flow, which is typically displayed in cubic feet per minute (cfm, cfm, ft³/min). These methods produce results that are typically accurate within +/- 3% or better.

Step 1: Flow bench data must first be converted to actual cubic feet per minute (acfm).

**acfm** = $\sqrt{[(29.92 \div P_b) \times ((T + 460) \div 520)]} \times F_b$ Flow

where acfm = actual cubic feet per minute, $P_b$ = local barometric pressure in in.Hg, T =°F, $F_b$ Flow = flow data from flow bench (cfm)

Step 2: The local air density must be calculated in lb. per cubic foot (lb/ft³).

$\rho$ = 1.325 x ($P_b \div$ (T + 460))
where $\rho$ = Greek letter rho (local air density in lb/ft³), $P_b$ = local barometric pressure in in.Hg, T = local air temperature in °F

Step 3: The flow data now in acfm is converted to mass flow in lb. per second.

$Q_m$ = ($\rho$ x acfm) $\div$ 60
where $Q_m$ = mass flow in lb/sec, $\rho$ = Greek letter rho (local air density in lb/ft³), acfm = actual cubic feet per minute

Step 4: The pressure ratio is calculated for later use. The pressure ratio should typically remain below 1.89 for the calculations to be effective and accurate. See the changes required when the Pr > 1.89 as an alternate calculation at Step 6 below.

$P_r = (P_b \div (P_b - T_p))$

where $P_r$ = pressure ratio, $P_b$ = local barometric pressure in in.Hg, $T_p$ = test pressure in in.Hg. 13.6 in.$H_2O$ = 1 in.Hg. Because this is a ratio, you can use any units as long as they are the same so that unit cancellation is accomplished.

Step 5: The air velocity through the device under testing must be calculated in reference to the Mach number.

$M = \sqrt{[(P_r \times 0.2857) - 1) \div 0.2]}$

where M = Mach number, $P_r$ = pressure ratio

Step 6: The final calculation for the effective flow area can now be accomplished. The effective flow area will always be less than the geometric flow area. The calculation will produce an equivalent flow area in square inches.

$A_e = EFA = [(Q_m \div P_b) \div \sqrt{(T + 460)} \times (0.91886 \times M) \times [(1 + 0.2) \times M^2)^{-3}]$

where $A_e$ = EFA = effective flow area, $Q_m$ = mass flow in lb/sec, $P_b$ = local barometric pressure in in.Hg, T = local air temperature in °F, M = mach number

If the $P_r$ is greater than 1.89, the formula above becomes:

$A_e = Q_m \sqrt{[(T + 460) \div (0.53175 \, P_b)]}$

However it is strongly suggested that testing be done at pressure ratios less than 1.89 for more consistent results.

Another methodology of calculation based on Fleigner's work for the effective flow area is:

$A_{eff} = Q_m \sqrt{(R \div 2_g)} \sqrt{(T) \div P_0} \, 1 \div \sqrt{(\Delta P \div P_0)}$

where $A_{eff}$ = effective flow area in square inches, $Q_m$ = mass flow air in lb/sec, R = universal gas constant 53.3 ft-lb in °R, g = gravitational constant (32.17 feet per second), T = temperature in plenum area in Rankine units (°F + 459.7, or rounded to 460), $\Delta P$ = pressure drop across device in pounds per square inch, $P_0$ = ambient barometric pressure in lb. per square inch. In order to simplify, since R and g are constants, that portion of the equation becomes:

$\sqrt{(53.3 \div 64.34)} = 0.9101711$

With accurate flow data and some reliable cross-sectional measurements of the port you could rate this port for efficiency based on the effective flow area. This particular port feeds a cylinder that is 75 cubic inches, yet there is still a need to keep the velocity up for cylinder filling.

In order to calculate the pressures in pounds per square inch (psi) when observed in in.Hg, divide inches of mercury by 2.035. In order to calculate the flow bench pressures in psi when the test pressure is in in.$H_2O$, divide the in.$H_2O$ by 27.67.

Another method using Fleigner's work as a baseline is the following equation:

$AC_d = W \div 1.0988 \sqrt{[(\Delta P \, P_{up}) \div T_{up}]}$

where A = square inches, $C_d$ = coefficient of discharge, W = airflow in lb/sec, $\Delta P$ = air pressure drop across part (psid), $P_{up}$ = upstream or external absolute pressure (psia), $T_{up}$ = upstream or external absolute temperature (°F + 460) Rankine.

It should be obvious that using the equation above one must either know the specific coefficient of discharge (Cd) or resolve it algebraically. The area times the Cd will equal the effective flow area. This method is more cumbersome than the earlier outlined methods.

Application of the effective flow area (EFA) calculations to cylinder head ports is a very handy way to evaluate the port without internal probing with a Pitot tube for local velocity numbers. Probing a port for local velocity is difficult to get very reliable numbers because the Pitot tube somewhat gets in its own way, and a very detailed map of the port velocity variation is even more difficult.

The effective flow area of a port system divided by the geometric flow area (covered in the measurement of port cross-sections) generates a coefficient of discharge ($C_d$) for the port system. Note that the number will change for any variation in the cross-section of the port. It is normally better to classify the port system at the point of smallest cross-section.

## Effective Flow Area Reference Examples

Without showing the complexities of the calculations, a flow bench reading of 400 cfm at 25 in.$H_2O$ at 29.92 in.Hg and 60°F would yield an EFA of 2.69 square inches. If the 423.2 was at 28 in.$H_2O$ (400 x 1.058), and the conditions were the same, the EFA = 2.664 square inches. And if the actual measured port cross-section (geometric flow area) was 3.25 square inches then the coefficient of discharge ($C_d$) would be 0.828 and 0.820 respectively. Both numbers would be efficient if they were real. Nicer to move toward 100% efficiency, though. Studying the efficiency of a port by using the effective flow area method is a very good way to help with analysis of the relationship of airflow numbers with the size of the port.

---

### FLOW WAND
*By Byron Wright, courtesy of Thorpe Engine Development*

These flow flags or wands are used to help identify where to measure the local air velocity with a Pitot tube. The flow wand is also used to help visualize three different things: stable airflow, turbulent airflow and airflow direction. As you move the wand around in the port it will encounter air that is moving very fast, slow, not at all, turbulent and very turbulent. And these will all have different effects on the string's reaction. The challenge is to interpret the string's motion and make appropriate changes to the shape or size of the port.

You are going to be observing the movements of the string as it encounters the different types of air. In areas of stable flow, the string will remain straight and tight. That is what you are aiming for: stable air in every part of the port. In areas of turbulence, the string will "fan out."

Start with a 1" length of string. With less extension the string's motions are not as pronounced, but will still show a lot of air types, including direction. With more string extended, the string reacts more, but with too much extension it can get in its own way. Extensions of 3/8 and up to 1 1/2" work well. Probe with 1", then 1/2", then with 1 1/2" and you will get a good understanding of where the air is and the type of air in the port.

Move slowly as you explore the port; find an area where the string stays straight and "tight." In this area, remember you need to think 3-D, move your string around in a zigzag pattern. Gradually make your "back and forth" and "up and down" motions bigger and bigger until the end of the string just begins to vibrate. What you are doing is determining the area of stable flow. You go just to the edge, where the string just starts to vibrate, to map out and determine the stable areas. Go slow, or you're going to miss spots, and don't wave the wand all around, you won't learn anything.

Be sure to do the whole port. Map out in your mind only one thing—the stable areas. As you move around the port, don't worry about the string waving all around, you're not concerned with those turbulent areas yet. First, find all the stable areas in the whole port. Keep some notes, for example: "bottom right side 5 o'clock, from mouth to 3" in, 1/2" high. Right behind push rod corner 9 o'clock, 4 1/2" in about 1/4" deep." One thing you might want to do is draw a top view of the port and mark in your drawing where the different areas are. Maybe you'll want to mark these areas with an "s" and the turbulent areas with a "t."

Next, find the turbulent areas. This is where it gets a little more complicated, because we're going to need to rate the turbulent areas. Go to an area where the string fans out. Just like before, slowly move the end of the wand in a zigzag pattern to map out the turbulent area. Only this time, the edges will be defined by the string becoming straight. In that turbulent area, how wide is the string fanning out? Is is a 1/4", 3/8", 1/2"? That is an area of "mild" turbulence. Or, is the string wildly going in all different directions? That is a very turbulent area. There are areas in the port that are going to be more turbulent than other areas. And you need to know where those are so you will have knowledge of the different types of air in the port. When you rate the turbulent areas you can use a scale like 1–5. With #1 being not that active, and 5 being things are really mixing it up. I use plus and minus to fine-tune the scale. Numbers 1–10 are, for me, harder to judge. One to five works well with pluses and minuses.

You can hold your wand very loosely between

your fingers and the wand will vibrate randomly as the air buffets it around. It's another input for you to determine the rating and strength of that turbulent area. Do the whole port and record it in your notebook, or on your drawing. Example: At 4" in and 1/3" up from floor to 1/4" from roof, about 1/2" wide @ 10 o'clock, rates a score of 4.

In the very turbulent areas you will see the string going from a horizontal direction to a somewhat vertical direction and switching from one to another randomly. Sometimes, in these very turbulent areas you might even see the string come back at you! Or the air might be taking it in a corkscrew motion. Depending on your position, the string will be switching back and forth, pointing toward the valve then toward the port wall or out to the port entry. Slowly pulling the wand in and out will pinpoint that area/position of instability. A common area is around the pushrod corner where the air is separating from the wall and the low pressure zone around the back side of the corner is trying to pull it back. Watch the string's reaction carefully. The flow direction and "types" of air are the information the string is telling you.

How much the string fans out is related to two factors; how large the turbulent area is and the amount of exposed string protruding from the end of the tube. Experiment with different string lengths and watch where the air takes the string. By having the correct amount of extension the string will fan out to a very close size of the turbulent envelope area. Being able to visualize the stable airflow, the turbulent airflow, and the degrees of turbulence really helps you to understand what the air is doing in the port. Now we need to ask some questions. Why is it stable here? Why is it turbulent here?

Turbulent areas are generally created by flow separation from the port wall. Why does the air leave the port wall? First remember air does not like to turn corners. It doesn't mind too much if it's going slow, but it still may leave the port wall, depending on the port shape. But generally speaking, the slower the air speed (velocity), the easier it will make the corner, if it is not too sharp. The higher the velocity, the more it dislikes to make a turn.

Using a flow bench is somewhat like taking a "snapshot" of a running engine. Your flow bench generates a constant velocity, which is NOT what the engine sees. The engine sees varying velocities over time. To a certain extent, testing at different depressions and test pressures will give you different snapshots of what is happening. Remember the formula $F = M \times A$? Acceleration is velocity x time. The flow bench doesn't deal with acceleration, only differential pressure, which produces a velocity, which is a function of orifice size and the magnitude of differential pressure.

If we substitute VxT for A, our formula looks like this:

**F = M x (VxT)**

For a given M (mass) and a given T, what would happen to F (force) as V gets bigger? What if V gets smaller? So the higher the velocity, the more force or energy it has, and the harder it will be to make the corner.

Here's something to think about: as M (mass) gets bigger, what happens to F (force)? As M gets smaller, what happens to F? Mass is a function of changing air temperatures and the varying volume and size of fuel droplets, depending on throttle position! What can we do to help our air/fuel particle turn around the corner? What if you made the corner 45°? You'd still fly off. How about 22 1/2°? You'd still fly off. What about 7 1/2°? You'd barely stay on the path. A 15° included angle is about the maximum diverging angle that air can handle and stay attached to the wall. What about a converging angle? Air will stay attached and can handle a larger angle here.

Here's another thing to think about: the difference in molecular weight between gasoline and alcohol and their stoichiometric ratios. If you're working on a port for gasoline or alcohol, would they be different?

What else could you do to get your air/fuel particle around the corner? You could slow down the flow and that would help. Roughing up the port in that area increases the boundary layer and helps the air stay attached longer.

Do you have a pushrod corner in your port? Do you know what that corner angle is? Now you know why there is a turbulent area off the apex of the corner with a dead area between it and the wall. Now it would seem that the right thing to do is straighten out the corner. Let me remind you about port volume and shape. Are you going to make the port like the Michelin man? That will not do. Shape is very important. Let's remember we are dealing with more than one thing here. One is quantity (we will need a certain amount to produce the predicted horsepower) and one is quality. If your port is changing velocities all over the place you might have the quantity but you certainly won't have the quality presented to the combustion chamber. The air and fuel won't be homogenous but spotty and dribbling into the combustion chamber—not good.

Unless you are designing a port from scratch, all you can do is move the port around a little bit, or fill in here or there within certain physical confines. The ideal port is one where in any position or area the air would be stable. But life is full of compromises, getting as close to ideal as we can is what it is all about. So don't feel like a failure if you haven't "created" a perfect port. Sometimes it is just physically impossible.

# Chapter 6
# Establishing Airflow Testing Standards

*Let us raise a standard to which the wise and honest can repair; the rest is in the hands of God.*
—George Washington

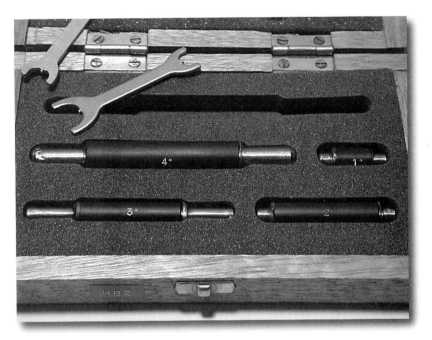

Standards for calibrating micrometers are accepted practice in metrology and most quality conscious machine shops. Flow benches need to calibrate to a known flow standard or base for reliable data and flow numbers. Those standards for airflow are called flow standards. Easy to say, harder to do.

The performance aftermarket doesn't currently use any universal flow testing standards. In fact, the OEM R&D departments don't seem to agree on test standards either. Each OEM has their own set of procedures and protocols, but there is nothing that is considered a true universal standard everyone can apply. Most have adapted to using 28 in.H2O test pressure or at least relate readily to the numbers collected at that test pressure.

However from time to time somebody pops up and sets a target for others to follow or dispute. So, now is a good opportunity to propose some test standards for flow benches in order to encourage some common ways to compare data.

Basically, I propose that all cylinder heads, manifolds, carburetors, throttle bodies, and various intake tract devices (excepting air filters) be tested in a standard manner that is described in detail. I also recommend that the same guidelines be applied to the exhaust side of the engine airflow path as well.

Because air filters have large areas and are essentially airflow straighteners or airflow smoothers, a different manner of testing is proposed.

## Airflow Straighteners

There are several types of airflow straighteners. One type is called a Zanker design and another is called a Sprenkle design. The Zanker type is grated with a disc of holes in one end. The holes have a certain pattern and size defined by specification, and the length of is equal to the diameter of the device.

The Sprenkle straightener is literally a group or bundle of tubes with each tube being equal to about 5% of the diameter of the total bundle. The bundle of tubes has to equal at least greater than 40% of the diameter for the whole bundle. Either device is called a flow conditioner. These type devices are used to follow a standard of flow measurement using orifices, nozzles, or venturi as determined by the ASME (American Society of Mechanical Engineers). These type flow conditioners are not to be confused with the same application of a laminar flow element, but they are very similar. The idea of these conditioners is to provide an orifice in the flow path by some type of airflow spoiler so the flow is smoothed but not completely laminar. Most orifice type flow benches use some kind of spoiler before the flow orifice(s).

The idea of a component flow-testing standard should not come as much of a surprise, because after all there are standards for testing and calibrating many devices that flow air. Perhaps applications of those standards set by the American Society of Mechanical Engineers, American Society of Heating and Refrigeration Engineers, or the Society of Automotive Engineers would be easy to put into service.

# Establishing Airflow Testing Standards

This UAV engine has been designed, tested and applied with airflow at the core of the program. Oil cooler, engine cooling, intercooler, and combustion air entries and exits have to fit in a confined space and produce adequate power. Proof testing was done at the US Air Force Academy's high altitude engine test facility.

The downstream side of an ASME sharp edge orifice is shown. The other side would be facing the airflow in a calibration verification test. Various sizes of these orifices are needed to properly calibrate a flow bench. This particular orifice is certified to flow within a few cfm of 400 cfm at 28 in.$H_2O$ test pressure.

Flow testing components should adhere to a standard form for test pressure measurement and airflow units. I propose that the standard test pressure be 28 in.$H_2O$ and the standard units for flow should be cfm (cubic feet per minute). This format would be referred to as *bench flow*. In my opinion, two modes of testing operations should be recognized as both separate and necessary.

1. Raw or as-tested on the flow bench with local atmospheric conditions listed but not initially used for anything except to accompany the flow data.
2. Corrected to known and accepted atmospheric standards if applicable.

The format for testing is outlined below. This is part of the overall proposal for attaining and applying common testing standards. All testing should be accomplished at 28 in.$H_2O$ test pressure and recorded or the actual test pressure should be recorded and the data should be corrected to 28 in.$H_2O$ test pressure and notated accordingly.

1. All inlet direction testing should involve a radius flow guide for testing the inlet direction.
2. The radius inlet guide should be a minimum of 1.5" thick and have radius of not less than 3/4". The material used for the guide can be anything; however clay should not be used because it is difficult to accurately duplicate the part from one test to another.
3. All exhaust direction testing (on cylinder heads) should incorporate an exhaust pipe simulation stub of at least 8" long with an inside diameter that is at least equal to the exhaust valve diameter. Regardless of the pipe diameter, a pipe should be used. The exhaust testing pipe can use a liberal radius bend if necessary.
4. A cylinder bore adapter should be used that has a bore diameter that is recorded with the test data. The length of the cylinder bore adapter should be not less than 4" and not greater than 6". The adapter's bore dimension should duplicate the engine's bore within +/− 1/32" (0.03125").
5. The cylinder head must be located over the cylinder bore adapter as it would be when attached to an engine. This is to ensure that the valves and combustion chamber are in the normal relationship with the bore.
6. A leakage test must be performed and results recorded for the test pressure selected (28 in.$H_2O$ standard).
7. A spark plug or a plug must be inserted into the spark plug hole and the presence recorded for the test. You would be surprised by how many tests are done without a spark plug inserted in the head. That mistake does improve the airflow numbers.
8. The valve diameter(s) and the stem diameters must be recorded.
9. The valve lift must be measured for each lift reference with a properly calibrated dial indicator. The amount of valve lift should be recorded at each data point. The data can be either at every 0.050" lift or at percentages of the valve diameter for a ratio of lift to diameter. If an L/D ratio is used, the lift increments will be at a minimum of 5%

# Engine Airflow

**Hand-shaping of intake and exhaust ports and combustion chambers takes a steady hand and considerable knowledge if the intended results are to be realized. Although Larry has a CNC setup, he still enjoys the challenges of making airflow work. Photo courtesy Endyn.**

changes. Either method must be recorded for easy comparison of data.

10. If a throttle body or carburetor is being tested, the test pressure chosen for the test is free. The test pressure used should be recorded, but the unit flow will be listed at an equivalent test pressure of 28 in.$H_2O$. See carburetor and throttle body testing descriptions in Chapter 10.

11. The local barometric pressure during the test should be recorded. The barometric pressure should be listed in in.Hg and it is preferred that the pressure reading be taken from a properly calibrated vertical mercury barometer. This is also referred to as "station pressure," which is a non-corrected pressure measurement. This is literally the pressure at the test site.

12. The local ambient air temperature during the test should be recorded. The temperature should be recorded in degrees Fahrenheit (°F).

13. The local atmospheric water content during the test should be recorded. The water content should be listed by reference to dry bulb and wet bulb temperatures. Relative Humidity reference is acceptable. A reference to dew point is optional.

14. The airflow bench should have reference to an approved calibration and the method should be described and recorded.

15. Accompanying the test data should be any notes concerning the test conditions or test procedures.

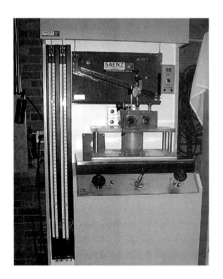
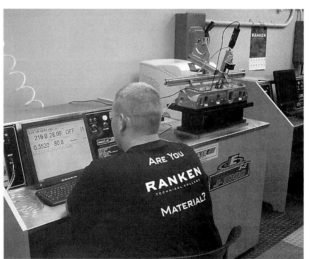

**A properly calibrated airflow bench is often the easiest way to measure airflow of components in order to compare results with other testers or developers. Good test results are often the result of applying good test procedures. Airflow benches of this general design are typically used for airflow applications from only a few cfm to over 2,500 cfm at test pressures from 3 in.$H_2O$ to more than 60 in.$H_2O$. Courtesy of Ranken Technical College.**

# Chapter 7
# Airflow Calibration & Measurement Standards

**This shows an easy to remember reference to what repeatability means. Not easy to achieve, but necessary for reliable data.**

*Therefore O students study mathematics and do not build without foundations.* —Leonardo da Vinci

## Calibration Standards

This subject is very sensitive to some people and not of much interest to others. The basic calibration of an airflow bench or a device to measure airflow through an engine or components is only important if the operator is interested in comparing data to another facility, textbook, or flow bench. Many folks (including some manufacturers of commercially produced flow benches) want to proclaim that their particular procedures and benches are accurate, but if the flow numbers are not referenced to a standard methodology it becomes a "he said, they said, we said" kind of reference and as a result is subject to a litany of criticism and disagreement.

*Repeatability*—The major issue for most operators working with an airflow bench is that it needs to be repeatable. An additional requirement is that the bench data should be accurate. The ideal circumstance would be for the flow results to be both accurate and repeatable.

Repeatability is defined as a test that readily duplicates another test under the same conditions. This is quite often interchanged with consistency of measurement. The repeatability of a test is an imperative when good test results are the goal. If a shooter's bullets were all bunched up in a very nice "group," but were 6" from the target center, then the efforts are very repeatable, but not very accurate. Another example is when five different people measure the same crankpin diameter with the same micrometer; they should be getting the same answer within +/- 0.0005" (one half of one thousandth of an inch). That would provide repeatable data, but says nothing about the specific accuracy of the measurement.

Accuracy is defined as a measurement that is referenced to a known and recognized standard.

This expression is quite often interchanged with the "right" or correct measurement. This term can be best described by considering that the center of a bull's-eye target is the accurate goal. How close the measurements are to the known accuracy standard would define the machine accuracy. Accuracy in flow numbers is verified by calibration to a known and recognized standard. The extended example of this measurement would be that the micrometer in the previous example was compared to a standard that was a known dimension such as the standards that are supplied with quality micrometers.

Accuracy and repeatability are the criteria that inspire confidence in the results of any test. But of the two, repeatability is the most desired. The capability of the machine to duplicate results gives the operator and the analyst more confidence that the tests are being done in the same manner. If one can add the element of accuracy to that package, then the goals are comfortable, achievable, and believable. This is sometimes referred to as a confidence interval and reflects a factor of reliability.

## Common Flow Measuring Standards

Most engineers reference one of three organizations when

**This shows a reference for what accuracy might look like on a target. The target is struck somewhat equally from the center, but not with strong repeatability.**

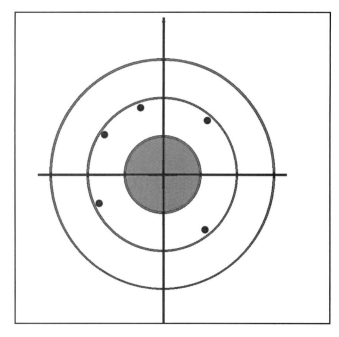

**What we really want in measurements or data. Accuracy with repeatability is a pleasure to deal with although not so easy to accomplish. For very dependable data you want to have both accuracy and repeatability. The confidence interval or confidence level is really what you are working for.**

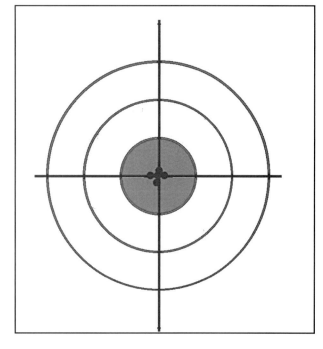

it comes to a standard of measurement: NIST (National Institute for Standards and Technology) or ASME (American Society of Mechanical Engineers) or SAE (Society of Automotive Engineers). But it is not uncommon for manufacturers or individuals to list some very imaginary capabilities for their test devices.

*NIST Standard*—NIST services for calibration of flow meters costs in excess of $4,700 per meter or orifice device and the flow data is provided with pressure, volume, temperature, and time. The NIST service claims to provide data that is true within an uncertainty of +/- 0.02% to +/- 0.13% (for dry air). The provided report would establish a Working Gas Flow Standard (WGFS). If the same testing was applied to establish a transfer standard, the cost is probably more expensive and it is charged at cost per time. This type of testing can be done on anything from sharp edge orifices or flat plate orifices to critical flow venturi to laminar flow elements (LFE), so if you wanted to verify the calibration of only three orifices that you had built the cost could easily go beyond $14,000 plus the freight and insurance. Perhaps there are some more practical ways to verify calibration before you want to check the calibration of the flow bench that you want to qualify.

If a manufacturer states that their equipment is traceable to NIST certification or some similar claim then you can ask to see a copy of the certification to see if they are blowing smoke. The certification form is small but it is official and it is illegal to claim a device is certified when it is not. It is also illegal to mark a device "patent pending" when it is not and that has been done before, too.

***ASME Standard***—ASME flow standards for flow benches are titled Measurement of Fluid Flow in Pipes Using Orifice, Nozzle, and Venturi (MFC-3M-1990 or MFC-3M-2004). So it is perhaps not very surprising that many commercial manufacturers of flow benches do not go to the extra trouble of providing this level of calibration.

This particular standard specifies the specific geometry and methods of use for the installation and flowing conditions for pressure differential devices. It is limited to orifice plates, nozzles, and venturi tubes when installed in a closed conduit and used to determine the flow rate of the fluid flowing in the flow path. The standard applies to pressure differential devices in which flow remains subsonic throughout the measuring section and where the fluid is considered as only single-phase flow. The standard is limited to applications of single-phase Newtonian fluid flow in which the flow can be free from pulsation effects.

The standard gives specific information for calculating flow rate and the associated uncertainty (data points) when each of these devices is used within specified limits of application (pipe size) and Reynolds number, which was discussed in Chapter 5. To review here, the Reynolds number is calculated from the following equation:

**NR = V x d x ρ ÷ μ**
**where NR = Reynolds number, V = average velocity of fluid, d = diameter of pipe, ρ = density of the fluid, μ = viscosity of fluid.**

# AIRFLOW CALIBRATION & MEASUREMENT STANDARDS

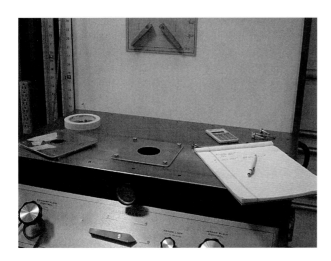

Verifying the calibration on a flow bench or completely calibrating a flow bench is not difficult if you have access to known and approved test orifices. The photo shows how to properly mount a calibration orifice on a flow bench.

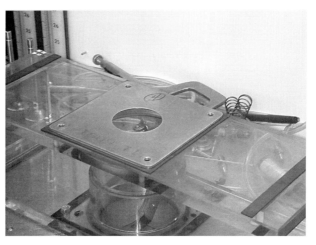

This is the improper way of mounting a calibration orifice because it does not follow correct procedures. The orifice mounted on a bore simulator will not give the same numbers as when mounted directly on the bench. Do not use bore simulators in this calibration or verification process because that is not correct procedure.

If the Reynolds number is less than 2,000 the flow is considered to be laminar. When the Reynolds number is about 2,700 to 4,000, the flow is understood to be laminar or in transition. If the Reynolds number is greater than 4,000 the flow is considered to be totally turbulent. The mention of the Reynolds number is for your basic understanding only and for a frame of reference concerning flow.

This described flow standard also covers flow meters that operate on the principle of a local change in flow velocity or flow parameters caused by meter geometry, resulting in a corresponding change or pressure between two set locations. There are several types of differential pressure meters in the marketplace and it is the purpose of this standard to address the applications of each meter and not to favor any particular meter or manufacturer.

The basic operating principles of a pressure differential flow meter are based on two physical laws: the conservation of energy and the conservation of mass. It is realized when changes in flow cross-sectional area and/or flow path result in a change or pressure. The differential pressure is a function of the flow velocity, the fluid path, and the measured fluid's properties. The scope of this particular standard includes devices for which adequate direct calibration experiments have been done. The tests should include enough repetitions for sufficient data coverage, in order to enable exceptionally valid systems of application to be based on their results and established flow coefficients. The data is to be provided with stated and known uncertainties. The devices installed in the flow stream are referred to as primary devices. Primary devices may also include the associated ancillary upstream and downstream piping. Other instruments required for the specific flow measurement are often referred to as secondary devices.

So as you can imagine the issue of calibrating to a standard is not as simple as it might seem. This is not an insurmountable process, but it is sometimes much more complex than it seems at first. Again the issues of knowing where the numbers come from is part of knowing what the numbers mean.

As an example, you can make your own test orifices and declare that their bench has been "calibrated with orifices" even if those orifices do not follow design procedures set by any standard. If you had an orifice that measures 2.50" in diameter,

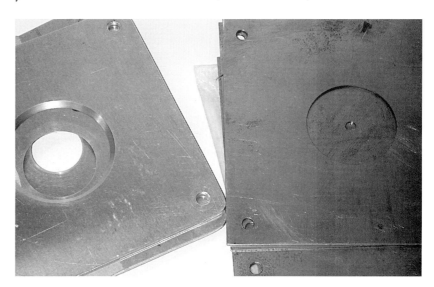

The orifices on the left are certified calibration orifices and the orifices on the right are not. Both will impose a load on a flow bench, but different numbers would result from those on the right. It doesn't mean they (orifice plates on the right) are not useful for some work, but calibration is not one of them. They would mostly be useful for simple comparison from one bench to another.

and stated that it flow-tested to equal some cfm at some differential pressure ($\Delta P$), the real answer can be all over the map. Even simple measurements become a problem.

## How Much Flow is Produced Through an Orifice?

The approved method of calculation that is described in the ASME specification and requires the calculation of many levels before a final flow number can be assigned to an orifice, nozzle, or a venturi. Of course it depends on several things and that starts with the physical shape of the orifice. It also depends on the $\Delta P$ or differential pressure across the orifice. It is worthwhile to note here that the normal measurement of the pressure across the orifice is done in inches of water and is typically measured with a manometer. Most vertical manometers are in the vicinity of +/– 1% in accuracy. Most inclined manometers are typically +/– 0.5% in accuracy.

I am inclined to believe that if liquid manometers are used that flow agreements with a standard should be no greater than 1% in deviation from the standard number for flow.

The orifice should have an edge that is not less than 0.005" thick and preferably less than 0.060" thick where the orifice is precisely bored through the plate material. If the plate material is thicker, the bottom of the orifice should be beveled at a 45-degree angle. The orifice should be used in such a way as to face the flow direction of flow with the sharp edge of the orifice. The edge of the orifice is not to have any scratches or nicks. There are many other requirements that are listed in applicable ASME specifications.

The basic equation for flow calculation that is required by ASME for flat-plate or square-edged orifices is the following:

$Q_m = 0.09970190 \times C \times Y \times d^2 \times (h_w \times \rho_f) \div (1 - \beta^4)$
where $Q_m$ = mass rate of flow in lbs/sec, C = coefficient of discharge, Y = expansion factor, d = diameter of orifice in inches, $h_w$ = differential pressure, in.$H_2O$, $\rho_f$ = Greek letter rho (density of flowing fluid), $\beta$ = diameter ratio

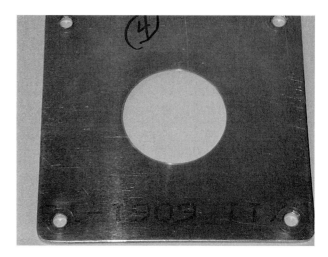

Basic calibration for a flow bench should rely on a known and traceable orifice or device such as these sharp edge orifices. The traceable orifice will come with a certification sheet and it takes more than one to properly calibrate a flow bench properly.

# Chapter 8
# Comparing Flow Numbers

The test pressure on most flow benches that use manometers is measured with a vertical manometer typically in inches of water (in.H$_2$0). The test pressure manometer is on the left side of this 600 cfm bench.

*Pleasure in the job puts perfection in the work.*
—Aristotle

Comparing flow numbers is a task anyone involved in flow testing must endure. Even if the number comparison is done on the same components and flow bench, it is important to know how to compare the numbers so that the time and effort is worthwhile. The comparison process is also a necessity to evaluate published numbers vs. your own developed flow numbers. The first thing that one learns in flow testing is that you must ask (or qualify) at what test pressure the flow numbers were recorded.

## Differential Pressure (Test Pressure, $\Delta P$)

The differential pressure across an orifice or across a component such as a cylinder head, manifold, throttle body, or carburetor needs to be accurately measured. The test pressure is normally measured on a liquid manometer or pressure transducer and is displayed in relative inches of water. Some test pressures are measured by using pressure transducers, and the test pressure is displayed in digital form after the data collection system has taken the initial pressure signal and manipulated it for display on some type of electronic display or system.

Even though most manufacturers of flow benches and some of the instruction manuals provide a chart for comparing flow numbers, there is little universal agreement on the flow numbers or the methods of comparison. The comparison of one test pressure vs. another can be done if you understand how to do so. But not everyone has a flow bench or even appropriate instruction, so the following information will help. With sensible application, you can compare the numbers if you know how the numbers were collected and where the numbers came from, a theme I keep repeating throughout this book

The charts in most instruction manuals for flow bench operation are based on the square root of the pressure ratio method. If you have flow numbers at a known test pressure and want to compare those numbers at a different test pressure, it is easy to do. The method is very standard for comparing differences in flow vs. pressure and is also in every textbook on both fluid mechanics, fluid dynamics and in most physics texts as well. So, it is based on principles of science that have been in around for hundreds of years.

*Example*—As an example, if you have flow numbers at 10 in.H$_2$O test pressure and would like to know what the flow should be at 28 in.H$_2$O test pressure, the process is:

$$\sqrt{(28 \div 10)} = 1.673$$

You would multiply the flow numbers taken at 10 in.H$_2$O test pressure by 1.673 to see what the flow should be at 28 in.H$_2$O. What this means is that if you have flow numbers at a test pressure of 10 in.H$_2$O and you are comparing with someone that has flow numbers at 28 in.H$_2$O, then you would only believe a little over 59.7% of what they are saying to you. That puts some of these numbers into a perspective that's easier to understand. More and more testing is being done these days at test pressures in excess of

# Engine Airflow

Higher quality flow benches using manometers utilize special snap lock connections so that the connections do not leak. Even small leaks will drastically affect flow readings. This shows both test pressure and spare manometer and flow manometer with these type connections.

Can you compare your numbers to another flow bench such as this computer controlled 600 cfm unit? Just a few things are required so you can do the simple math if required for an easy comparison.

28 in.$H_2O$. Most testing today is done between 28 in.$H_2O$ and 60 in.$H_2O$. There are still some holdouts that test at lower test pressures either because of inadequate power to move the air or because their flow bench was built too small in flow capacity.

*Comparing Flow Numbers at Different Test Pressures*—The comparison of flow numbers at different test pressures is easy once it is explained. The method is the "square root of the pressure ratio," which yields a multiplier to use for the easy comparison of flow numbers at various test pressures. In the most common format, typically one wants to compare flow numbers in cfm at one test pressure to cfm at another test pressure. Flow data without a reference to the test pressure at which it was measured is virtually useless.

The equation for calculation is:

$$Q = \sqrt{(P_1 \div P_2)} \times q$$

where Q = resultant flow, $P_1$, $P_2$ are pressures for test and comparison, q = original flow number.

The chart on the next page is provided for a quick reference for comparing airflow numbers at different test pressures. It won't take you too long to realize the relationships of test pressure and resultant flow. It becomes almost a given that you ask what test pressure flow data was taken at just to put it all into some usable perspective.

Note that for the chart, the calculations have been done as described, but the multipliers displayed for the chart are rounded off. For a higher level of accuracy, apply the method that has been outlined and see if it adds appreciable data for comparison. If you want to compare other numbers at other test pressures, you can use the simple equation described previously to obtain a multiplier. It should be obvious that if you have collected data at a higher test pressure number than you want to compare to a lower number, that the multiplier will be less than 1. If you want to compare potential flow numbers at higher test pressures than the data was collected originally, then the factor will be greater than 1.

The top of the chart on the next page is referenced in in.$H_2O$ at which test data was measured and recorded. The left side of the chart is various test pressures that could be used to compare the collected data.

The chart has a reference for 20.4 in.$H_2O$. That reference is because 20.4 in.$H_2O$ is equal to 1.5 in.Hg and that reference is also where many four-barrel carburetors have been rated in the performance and aftermarket industry. That number and methods of carburetor and injector throttle body testing and comparison will be discussed in detail in other chapters.

Study the chart and use your own calculator to see how the numbers in the chart were calculated. Only trust numbers that you know how they were calculated.

As an example, if you have test data collected at 28 in.$H_2O$ test pressure and you wanted to compare it to potential results at only 10 in.$H_2O$, then you would multiply your original data in cfm by 0.60 to get the new number. This multiplier is found at the intersection of 28 on the top and 10 on the side. So if you had a flow number collected at 28 in.$H_2O$ that was 235 cfm, another person might expect to collect data of only 141 cfm at 10 in.$H_2O$ test pressure.

You can even use the above method for checking potential flow changes in fuel pressure to predict the change in fuel flow because the multiplier is dimensionless. As long as the units to be compared are the same, the square root of the pressure ratio is very convenient and easy to use.

# Comparing Flow Numbers

## Original Flow Data Taken at These Test Pressures (multipliers rounded)

| Compare At these test pressures | 10 | 15 | 20 | 20.4 | 25 | 28 | 30 | 35 | 40 | 45 | 50 | 55 | 60 |
|---|---|---|---|---|---|---|---|---|---|---|---|---|---|
| 10   | 1.00 | .81  | .71  | .70  | .63  | .60  | .58  | .53  | .50  | .47  | .45  | .43  | .41 |
| 15   | 1.23 | 1.00 | .87  | .86  | .77  | .73  | .71  | .65  | .61  | .58  | .55  | .52  | .50 |
| 20   | 1.41 | 1.22 | 1.00 | .99  | .89  | .85  | .82  | .76  | .71  | .67  | .63  | .60  | .58 |
| 20.4 | 1.43 | 1.17 | 1.01 | 1.00 | .90  | .85  | .83  | .76  | .71  | .67  | .64  | .61  | .58 |
| 25   | 1.58 | 1.29 | 1.12 | 1.11 | 1.00 | .96  | .91  | .85  | .79  | .75  | .71  | .67  | .65 |
| 28   | 1.67 | 1.37 | 1.18 | 1.17 | 1.06 | 1.00 | .97  | .89  | .84  | .79  | .75  | .71  | .68 |
| 30   | 1.73 | 1.41 | 1.22 | 1.21 | 1.10 | 1.04 | 1.00 | .93  | .87  | .82  | .77  | .74  | .71 |
| 35   | 1.87 | 1.53 | 1.32 | 1.31 | 1.18 | 1.12 | 1.08 | 1.00 | .94  | .88  | .84  | .80  | .76 |
| 40   | 2.00 | 1.63 | 1.41 | 1.40 | 1.26 | 1.20 | 1.15 | 1.07 | 1.00 | .94  | .89  | .85  | .82 |
| 45   | 2.12 | 1.73 | 1.50 | 1.49 | 1.34 | 1.27 | 1.22 | 1.13 | 1.06 | 1.00 | .95  | .90  | .87 |
| 50   | 2.24 | 1.83 | 1.58 | 1.57 | 1.41 | 1.34 | 1.29 | 1.20 | 1.12 | 1.05 | 1.00 | .95  | .91 |
| 55   | 2.35 | 1.92 | 1.66 | 1.64 | 1.48 | 1.40 | 1.35 | 1.25 | 1.17 | 1.11 | 1.05 | 1.00 | .96 |
| 60   | 2.45 | 2.00 | 1.73 | 1.72 | 1.55 | 1.46 | 1.41 | 1.31 | 1.22 | 1.16 | 1.10 | 1.04 | 1.00 |

The comparison of airflow data at different test pressures is easy to do using this chart. The chart provides you with the appropriate multipliers to use with your handheld calculator. For reference: 13.6 in.$H_2O$ = 1 inch Hg (mercury) = 0.49 psi = 3.386 kpa kilopascals = 0.034 bar.

## Keeping Numbers Honest

If you ever work with an airflow bench you will want to compare your findings with another site or another bench to make sure you are getting consistent numbers.

One of the easiest ways to make a simple comparison is to use a common flow test orifice. The test orifice should be either a flat plate orifice (very easy to machine) or a sharp edge orifice. The reason to use some device such as described is that it takes away many variables if you attempted to use a cylinder head and it also removes the emotions from a simple technical solution.

What size orifice should be considered? Remember this device for test comparison is not a calibration device, just a test device. It is not designed to test the full range of any bench, but would be for a comparison at the same test pressures on two or more benches and test sites. Flat-plate orifices and sharp-edged orifices are very predictable in what their flow characteristics are and as a result make great tools with which to compare various benches and sites. I recommend that you use a simple standard flat plate orifice to compare various flow benches in order to see how much variation in data is apparent early on. This type of orifice is less expensive to fabricate and if you use one with a diameter of about 1.875", then it should flow close to 250 cfm at 28 in.$H_2O$.

What if there is more than one orifice in a series? First there must be some assumptions about the application so that we are dealing with a perfect gas

## Coefficients for Estimating Horsepower

| Test Pressure | $C_{pwr}$ | 8 cyl. | 6 cyl. | 4 cyl. | 2 cyl. |
|---|---|---|---|---|---|
| 3 in.$H_2O$ | 0.787 | 6.30 | 4.72 | 3.15 | 1.58 |
| 5 in.$H_2O$ | 0.608 | 4.86 | 3.65 | 2.43 | 1.22 |
| *10 in.$H_2O$ | 0.430 | 3.44 | 2.58 | 1.72 | 0.86 |
| 15 in.$H_2O$ | 0.350 | 2.80 | 2.1 | 1.40 | 0.70 |
| 20 in.$H_2O$ | 0.304 | 2.43 | 1.82 | 1.22 | 0.61 |
| 20.4 in.$H_2O$ | 0.301 | 2.41 | 1.81 | 1.21 | 0.60 |
| 25 in.$H_2O$ | 0.272 | 2.18 | 1.63 | 1.09 | 0.55 |
| 28 in.$H_2O$ | 0.257 | 2.06 | 1.54 | 1.03 | 0.52 |
| 30 in.$H_2O$ | 0.248 | 1.98 | 1.49 | 0.99 | 0.50 |
| 35 in.$H_2O$ | 0.230 | 1.84 | 1.38 | 0.92 | 0.46 |
| 40 in.$H_2O$ | 0.215 | 1.72 | 1.29 | 0.86 | 0.43 |
| 45 in.$H_2O$ | 0.203 | 1.62 | 1.22 | 0.81 | 0.41 |
| 50 in.$H_2O$ | 0.192 | 1.54 | 1.15 | 0.77 | 0.39 |
| 55 in.$H_2O$ | 0.183 | 1.46 | 1.10 | 0.73 | 0.37 |
| 60 in.$H_2O$ | 0.176 | 1.41 | 1.06 | 0.71 | 0.36 |

*$C_{pwr}$ for 10 in.$H_2O$ = approx 0.43. On today's well-developed engines, this might even be 0.45 or even 0.46, but that is not really important. For comparisons use the same numbers in estimating.

*Test Comparison Procedure*—When comparing flow numbers, it is imperative that you use the same methods of testing and the same components. The cylinder head (or other device to be compared), radius inlet guide, exhaust pipe section, cylinder head bore adapter, locating devices, and dial indicator and valve opening attachment are all necessary parts to accomplish the testing. At the same time, the same test pressure references and the correct data for calibration or equalization need to be used. If these simple guidelines cannot be adhered to for any reason, then the direct comparison is not very reliable.

## Calculating Horsepower Based on Flow Bench Data

Various performance coefficients that have been used over the years were developed and are based on some very accurate empirical data for estimating power and other parameters. Even more than 50 years after their introduction to the marketplace they are still a good indicator of the power capability of an engine solely based on its airflow capacity. Today's engines have become more efficient because of airflow improvements that have been generated by thousands of people searching for more power. At the same time that airflow has improved in engines, the available quality of pump fuel has decreased. The prediction of horsepower based on airflow numbers can be applied if the test pressure is known. The results are a good estimate of the engine's capacity to produce power if everything in the system is optimized to take advantage of the airflow available. An accurate estimate of the power capacity of the engine is dependent upon having accurate flow numbers for the complete airflow system, including the cylinder head, manifold, carburetor, or fuel injection system.

You can come up with your own power predictions based on reliable dyno and flow bench data. After enough data is gathered, you can begin to see a trend of relationships between measured airflow of components and the power the engine produces as a result of using the airflow. Then you would have your own empirical database from which to draw conclusions.

Or you can apply some sensible logic for estimates that are close enough for planning and applications that involve components that affect the engine's airflow.

The horsepower coefficient varies with the test pressure and the chart above can be used for a quick evaluation of airflow numbers at different test pressures. Note that the airflow numbers are for a complete flow path, including the cylinder head,

(air) and that the velocity prior to the second orifice in the flow path is going to be low. If those criteria are met, then the following calculations would work:

$$M = C_1 \times A_1 \times \rho_1 \times a_1 \times \Phi_1 \times (p_2 \div p_1)$$
$$= C_2 \times A_2 \times \rho_2 \times a_2 \times \Phi_1 \times (p_3 \div p_2)$$

where C = orifice coefficient, A = area of orifice, $\rho$ = density, a = sonic velocity, $\Phi_1$ = compressible flow function

So what you end up with is a ratio of flows of one orifice versus the next. The sonic velocity in feet per second for air is:

$$a = 49 \times \sqrt{T}$$

where T = °F + 460 and the value of
$$\Phi_1 = (2 \div k - 1) \times [(p_2 \div p_{01})^{2/k} - (p_2 \div p_{01})^{k+1/k}]$$

What all this shuffling of numbers really means is that if two orifices are in a series where one orifice is at the entry of a box-type plenum and the second orifice is at the exit of the plenum then there is a ratio of pressures that will predict the flow differential. That is probably more than you wanted to know about airflow in conditions that are not directly related to a specific engine part, but at least you have been exposed to some of the oddities of how airflow works in the context of measurements.

## Coefficients for RPM

These numbers show at which rpm peak power should occur based on measured airflow.

| | |
|---|---|
| 3 in.$H_2O$ | 3453 |
| 5 in.$H_2O$ | 2828 |
| 10 in.$H_2O$ | 2000 |
| 15 in.$H_2O$ | 1633 |
| 20 in.$H_2O$ | 1414 |
| 20.4 in.$H_2O$ | 1400 |
| 25 in.$H_2O$ | 1265 |
| 28 in.$H_2O$ | 1196 |
| 30 in.$H_2O$ | 1155 |
| 35 in.$H_2O$ | 1069 |
| 40 in.$H_2O$ | 1000 |
| 45 in.$H_2O$ | 943 |
| 50 in.$H_2O$ | 894 |
| 55 in.$H_2O$ | 853 |
| 60 in.$H_2O$ | 817 |

When you get used to manipulating the numbers it is easier to make some fairly accurate predictions based on airflow. This welded H-D head is reworked to provide much better flow characteristics without raising the rpm potential beyond the capability of the engine. Photo courtesy 10 Litre Performance.

manifold, throttle body (or carburetor) for best results. It takes more time to have complete measurements, but they provide a more accurate prediction of the potential results. The equation:

hp/cyl = $C_{pwr}$ x test flow
where $C_{pwr}$ = Coefficient of power, test flow = cfm flow at the same test pressure that the $C_{pwr}$ is applied.

*Don't Out-Trick Yourself*—A reminder: The numbers in the chart above are empirically derived and automatically assume that the engine is using gasoline for fuel. Empirical means information derived from experiment and observation rather than theory and is based on practical experience or deduction. And being empirically based also means that the numbers might be constantly changing because the other numbers that support the conclusion or deduction are also changing. As dynamometer data and as flow bench data has become more accurate and repeatable (hopefully) the empiricism gets narrower and more finite.

**Example:** If you have system airflow numbers recorded at 28 in.$H_2O$ and the flow was 200 cfm, then the calculation would be:

hp/cyl = 0.26 x 200 = 52 hp/cyl

If you were working on an eight-cylinder engine, then 8 x 52 = 416 hp capacity. This of course assumes that each of the ports flow the same number. More accurate results can be applied if the same calculation is done for each port if the airflow is not the same. It is fruitless to measure just one port/runner/manifold combination and assume that all will flow the same. It is not uncommon for there to be at least a 10% variation and sometimes more across the parts. It is advantageous to make them much closer in capability with careful shaping.

The potential for power production is often there by providing enough airflow and fuel flow for combustion, but if the engine is using the incorrect camshaft and exhaust system or if it has a poor ring seal or poor sealing valve job then the engine will fall short of the predicted power level.

## Estimation of Peak Power Rpm Based on Airflow Numbers

The prediction of the rpm at which peak power will occur (based on airflow) is an additional useful way to evaluate airflow numbers. It is also an easy way to see the effects of changing engine displacement. The equation:

rpm at peak power = [($C_{rpm}$) ÷ (displacement/cyl)] x cfm]
where $C_{rpm}$ = Coefficient for peak power rpm calculation, displacement/cyl = displacement per cylinder in cubic inches, cfm = cubic feet per minute from flow bench data taken at a given test pressure.

Those empirically generated numbers in the chart above also assume that the engine is using gasoline for fuel and are empirically derived. Empirical methods are sometimes more valuable to the observer or tester than to the theorist because of constant testing and comparison of results produce recognizable trends. It is also faster to reach a usable solution when using this method of investigation.

**Example:** Using the same numbers of 52 hp/cylinder that were applied in the previous

# Engine Airflow

There are many reasons to test on a dynamometer and one of them is to verify the rpm at which peak power is made and to establish the power curve. Of course tuning is also a plus. Photo courtesy Dynamic Test Systems.

example above, 200 cfm at 28 in.$H_2O$, and the engine is an eight-cylinder that has a displacement of 355 cubic inches. Thus, 355 ÷ 8 = 44.375 cubic inches per cylinder. Applying the equation and solving for rpm at peak power, where:

$rpm_{pp}$ = (1196 ÷ 44.375) x 200 = 5390 rpm

Even though the valvetrain might have the capacity to increase the rpm to 7500 or 8000, it would not be useful to do so as the engine literally runs out of air.

Just for fun, what would happen to this number if the engine were 455 cubic inches? Now the engine displacement divided by the number of cylinders yields an entirely different number. So, 455 ÷ 8 = 56.875. Applying the equation for rpm at peak power:

$rpm_{pp}$ = (1196 ÷ 56.875) x 200 = 4206 rpm

It is not uncommon to turn the engine rpm to about 8% to 10% or so past the peak power rpm to allow for gear ratio changes on gear shifts. The preferred exact shift points will vary with gear ratio spreads or torque converter characteristics and rate of acceleration of the vehicle.

There are many applications that involve a specific need to know the airflow of engine components and how the engine uses air. Applying some simple equations can help to compare the relative airflow of components.

There are many relationships that can be enhanced if the airflow is known, including valve timing (camshaft selection), inertia tuning factors of intake, power per cylinder capacity, and the rpm at which peak power will occur.

There are some very interesting things that can be learned from the study of airflow through engine components and the engine itself. Because the engine is a self-driven air pump, many of the characteristics and its capability to produce power and the combustion of the engine are set by its capacity to flow air.

What happens if you have tested all the components in the airflow path except the air cleaner housing and the air cleaner housing has a flow restriction from the way it is designed? The engine will suffer as a result of the inferior or poorly matched components. That is why it is important to consider everything in the complete airflow path.

However after all those words, there are some other things that come into play that have a controlling effect on the rpm at which peak power will occur. One that is very significant and you should be aware of is the cross-section of the port and how it has an influence on the characteristics of the engine's torque and power curves.

# Chapter 9
# Flow Testing Intake & Exhaust Components

This might not look like the controls of a racer but you can race it if everyone in the race follows the same rules and procedures. Variations in flow benches can generally be explained to those that will listen to logic. Ya wanna race?

*In science there is only physics; all the rest is stamp collecting.*
—Lord Kelvin

## Flow Testing Cylinder Heads

A very good guideline for the study of airflow through engines is that shape is more important than size. Shape and direction of airflow of a port often helps the characteristics of the engine's capability to produce good torque and horsepower numbers. A well-shaped port that flows well will even have a softer or quieter sound when being tested on a flow bench. Harsh sounds of the airflow path are often the result of very turbulent or perhaps detached flow. Detached flow leads to a phenomenon known as bi-stable flow and sounds a lot like bed sheets flapping in the wind or the random suction sounds of a dentist's liquid suction tube in your mouth. Either way, if the airflow sounds ugly, it probably is. Irregular flow also has a sound associated with it. It is a sound that somewhat pulses but not on a regular time basis. It might even be described as the random sounds of smooth and not so smooth airflow. This can also be of the bi-stable family. There is no simple medication for a bipolar (bi-stable) port and it takes work to resolve the problem of confused airflow.

About the last thing that you want to do is pick up a grinder before you understand what a particular cylinder head is doing with the airflow. Sometimes (too often) it is very worthwhile to see what happens if a port or runner was filled in some places (modeling clay is great for this type of testing). Just because the port is a given size doesn't mean that it was from some wonderful design. It might have been designed by a committee or was the result of some casting

This port and cylinder head cutaway shows how there are many places in the port for air to get confused but not half as confused as a beginner with a grinder in his hands. Shape is much more important than size. Courtesy Meaux Racing Heads.

compromises made in the foundry process. It might have even been just too big or too crooked to start with. The reason for testing is to learn how to improve the component being tested or designed. Shape, not size, is one of the most critical considerations in ports and runners.

# Engine Airflow

This is a cutaway of an inline six cylinder head. Note where the coolant passages are. A common intake port with divider poses some interesting problems that can be overcome with a little planning.

This close-up of a shaped port has been rough cut with a carbide cutter on the cast iron head. There is not a lot of performance to be gained by an extra smooth surface beyond 80-grit or so, but it does look nice.

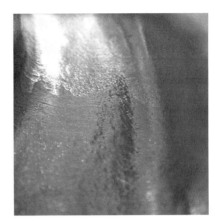

An old exhaust valve is used to prevent damage to the seat area. This guy is working on the exhaust side of the combustion chamber and will eventually shape the complete chamber with an old intake valve in place, too.

This is the intake side of a three-valve head that has had only a minor amount of reshaping. The intake ports are bifurcated (divided into two branches) and have a very slight short-side radius. Cautious flow work paid off in increased power.

The surface finish of the port walls is not important until you have the opportunity to study the effects of surface finishes on airflow. Again, a well-shaped port will flow better regardless of how smooth or rough the port walls are. Normally speaking, the finish of the cylinder head port walls is more than smooth enough if finished with a 36-grit abrasive roll. There is generally no gain at all in flow from making the surface finish any finer than with an 80-grit abrasive roll. However, the surface finishes that are often sold with aftermarket heads are in the 120- to 250-grit range, and although pretty and shiny, the finish does not add to the airflow performance of the port. Sometimes appearances sell better than do the results. Sometimes marketing and sales hype outweigh the logic of form and function.

## Effects of Cylinder Head Design

Because airflow testing is not brand loyal, it makes no difference if a cylinder head is made for Chevrolets, Fords, Chryslers or Dodges. The airflow does not care if the head is from a Cummins or Caterpillar or Lamborghini or a Ferrari or an Offenhauser. The measurements will even work for a Kawasaki, Harley-Davidson, Yamaha, Suzuki, Honda, or even Ducati motorcycle engine components. Some of the higher end motorcycle engines have ports that are so straight that there is virtually no short-side radius, and the casting techniques are such that the aluminum is nicely done right out of the foundry. However, all ports can be tweaked a bit to improve the flow. Sometimes it is a little shaping of the seats (angles), valves and the combustion chamber to prevent shrouding the airflow as it exits the intake valve into the chamber that will make a difference.

The cylinder head design and engine configuration are considered to be boundary conditions as far as flow measurement is concerned. All that is important is the head's resistance to flow air. The flow bench provides a way to compare data of different components and modifications to determine which ones work, and which ones don't.

It is often worthwhile to study the cylinder head very carefully. How many bends or obstructions are in the port flow path? How much of an offset does

This cutaway section of Pontiac cast iron is a necessary study before beginning work in creating a SuperStock piece. Photo courtesy Meaux Racing Heads.

This shows the roughing in of a small block Chevy iron head. The area that seems like a large protrusion into the port is a boss around a head bolt hole. When the ports are finished the area will be reshaped so the ports will be tangential to the valve bowls without breaking through the boss. Learn to think like air.

the port centerline have with the centerline of the valve? Does the port have a nicely developed approach to the bowl area? Is the transition from the short-side radius gradual or rapid? Is the depth of the port from the valve seat to the short-side radius very long or is it short? How close to the combustion chamber wall does the intake valve get at maximum lift? All these questions are things to consider long before you do any grinding or machining. It is sort of like the old saying for measuring and cutting that states "measure often and cut once." That is much better than saying "I've cut it twice and it is still too short."

## Port Configurations

The ideal shape for an intake port would have no bends or turns in it. In the real world, ports must share some space with head and manifold bolts, valve guides and other components or passages. Surprisingly, the easy places to get to reshape or modify are not generally the areas that make the most difference in airflow. Some very important areas are the sections of the port that are just before and just after the valve seat. The area of the port that is known as the bowl area is very influential in shaping the airflow to go past the valve and seat as the flow enters the combustion chamber and the cylinder.

*The Port Entry*—The entry of an intake port in a cylinder head can be almost any shape, but the most common is a rectangular type shape. Recent late-model pushrod engines use a modified rectangle with a church window top because of allowances for port fuel injectors. The entry is typically a continuation of the runner from the manifold, and if it is a pushrod engine, it might have a pinch alongside the port to accommodate the pushrod. Intuitively this area is a restriction to airflow, but in reality it is not that restrictive when compared to other locations in the port. Traditionally the rectangular shape of the port has the longer (taller) portion of the port in the vertical location while the width is the more narrow position. Surprisingly the port entry at the cylinder head mounting flange is not typically the smallest cross-section area. There is much to be said for the more oval shape of an intake port having fewer problems of turbulence than in the corners of a rectangular port. Certainly the larger radiuses of a well-shaped rectangular port are beneficial, so the natural outgrowth in shape would be an oval shaped port.

*The Short-Turn Radius*—This is a description of the area of the port where a port might bend just before entering into the bowl area or might describe the floor of an exhaust port. The turn of the port is better if the port is comparatively straight (very little bend) so that there is little tendency for the flow to separate or detach from the short-turn radius. Separation is the phenomenon that refers to when the airflow is no longer attached to the wall or floor of the port. Think of it as a waterfall-like effect where the water cannot follow the confines of the stream floor, and it leaves the path to become a waterfall. The airflow behaves in a similar fashion. The short-turn radius is a very active portion of the port and should be shaped very gently so that separation or detachment does not occur. If the flow is attached, it has less internal drag; if it is detached or separated, it has a lot more drag and as a result will lose flow. Many ports operate with detached flow along the short-side radius. You can gain power by making sure the flow reattaches itself, using several different methods. Usually these methods include bumps in the short-turn path, or sometimes step-cuts or little ditches across the short-turn radius floor at 90 degrees to the airflow

# Engine Airflow

The finger is pointing toward the short side radius. It is imperative the flow stays attached to this region or the flow will separate and not use the capability of the port and valve combination.

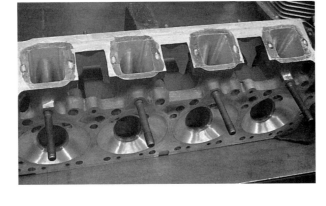

This 426 hemi head shows how the port centerline is in line with the valve and it is also aligned with the exhaust valve. During overlap, flow is very apt to exit right out the exhaust so camshaft selection is critical on these things.

This cross-cut photo shows what an ideal intake port shape might look like. The Kawasaki engine is one that uses four valves per cylinder. Notice there is almost no short side radius so little opportunity for the flow to become detached as it turns the corner into the cylinder.

path. You need to experiment with various methods to see what works to gain control of detached flow.

*Port Centerline*—The port centerline is the center of the port where the heart of the airflow would follow. Sometimes a port is positioned so that the centerline of the valve is not in line with the port opening and sometimes the valve is even tilted a few or several degrees as well. All those things have an effect on the airflow path. The more that the intake valve is inclined toward the airflow in the port, the more efficiently the air can envelope the valve (if things are nicely shaped). On the exhaust side, if the valve is tilted toward the exhaust side flange of the port, it is normally more efficient than if it was at 90 degrees to the port centerline of flow. The inclination of the valves to the port is sometimes a way to identify the port/cylinder head. Such as a 23-degree Chevrolet head is named that for the valve inclination, as are heads that are 15-degree and 18-degree configurations. That is not a universal way to identify cylinder heads however. Sometimes it is much better to identify the cylinder heads by a casting or part number, and then measure the pertinent things about the cylinder head. After all, when you can measure it, you can relate those measurements to more important things about the engine or components.

As an example, big-block Chevy heads are almost as diverse, but are not generally referred to as any particular degree of the intake valve inclination. Mostly because the normal valve and cylinder-head configuration is actually inclined in two planes, not just one. Thus the description of "canted valve" heads. So, beware that sometimes the slang terms don't really mean that much. In the final analysis it is what the cylinder head and related components do for the power curve that makes a much more important statement.

The guys that work with motorcycle engines have been spoiled for years because the basic engine designs are much more no-compromise type designs than their automotive counterparts. Some of the later design motorcycle engines have such beautifully shaped intake ports that they almost have no short-side radius to worry about. Yes, such designs often require long valve stems, but the stems are only 3.5mm (0.137") to 4mm (0.156") in diameter. Surely these are the high-end race bike types and use two intakes and two exhausts (not necessarily with the same stem size). Even the production street bikes are a study in all that is wonderful for airflow. There is something to be learned by looking closely at those engines that have been designed with very few compromises for airflow.

## Port Exit

All ports have points of entry and exit. The intake port exits into the combustion chamber and much of the loss in flow is during the rapid expansion and what is referred to as *exit losses*. That is why so much effort should be put into shaping the valve

# Flow Testing Intake & Exhaust Components

These simple pipes make the exhaust flow measurements on a flow bench more indicative of what happens on a running engine. They can actually improve the flow of the exhaust and change the ratio of the exhaust to the intake flow numbers.

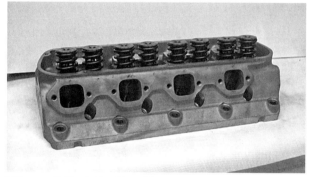

This small-block Ford head has evenly spaced exhaust ports. The reshaping must have worked well because the engine makes more than 700 hp. Photo courtesy Cathy Bevers.

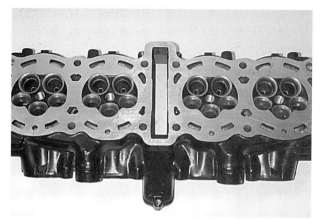

Yamaha did this five-valve design for their Genesis series in about 1984. Three intakes and two exhaust valves.

This is an odd one, a three valve. It is a Honda motorcycle engine that uses two intake valves and one exhaust.

seats, valve shapes and angles, and the combustion chamber walls of the cylinder head. Gradual changes in cross-sectional area are more supportive of fewer exit losses than the rapid changes. Think like air and run like the wind.

An exhaust port leaves the cylinder head, but it has to be connected to an exhaust pipe. Exhaust gases eventually exit through either a collector or tailpipe, cat converter (for street), then into the atmosphere via a muffler. The exhaust pipe is just a continuation of the exhaust port made from different material. Quite often, flow gains are discovered by reducing the amount of the exit losses from sudden expansion or directional changes. Therefore, if you are testing on a flow bench, you should always use a short length of pipe on the exhaust port for best results. This simple bit of advice will save you a lot of grief.

## How Many Valves?

The testing procedures vary a bit with the number of valves involved. On multi-valve engines, it is sometimes worthwhile to flow each valve individually to see what effect that particular valve has on the whole system. If an intake valve is off center with the stem of the valve, there will be an induced swirl that is often at the cost of optimum flow volume. It is common that one, two, or three intake valves are used per cylinder. While one or two exhaust valves appear to be the most common. The Yamaha Genesis motorcycle cylinder head that used five valves (three intake and two exhausts) was a drastic departure from other competitors. Honda engineers built an eight-valve (four intake and four exhaust) cylinder head for an oval-shaped piston design, but it never saw production.

*Hemi Configurations*—Most hemispherical configurations have the port centerline aligned with the valve stem centerline and as a result have very little induced swirl. Therefore, they flow pretty well and very easily during the overlap event (the period that both intake and exhaust valves are off the seat), and are prone to toss fresh fuel and air mixture right out the exhaust port. Pent-roof combustion chambers either have individual ports or paired

# Engine Airflow

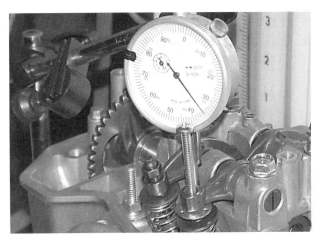

The tester of this Honda motorcycle engine had to get a little creative and chose to use the simple hardware approach for opening the valves for flow testing.

Although you might be able to pick up an old seat grinding set like this Kwik-Way SGF, you need to consider that it is not easy to properly use this kind of equipment. Sometimes it is best to leave grinding to the professionals, but you should have a thorough understanding of the process involved.

The shaping of the area just below and above the valve seat is critical to helping the airflow into the chamber area. The valve seat itself is critical for efficient sealing and life of the valve and the guide.

ports that might add to some swirl-induced problems. Most of the higher flowing pent-roof designs have a fairly flat and shallow combustion chamber. Many of the cylinder head designs that are used for diesels have ramps or drastically offset valves to the port so that swirl motion is greatly enhanced. The so-called high swirl ports can be generally improved in airflow rating by removing some or all of the induced swirl characteristics. Every time that the airflow has to change directions (even slightly) it costs some airflow. There is not very much reliable evidence that swirl produces better power than filling the cylinder in other ways (such as via the direction of tumble). Higher swirl ports are often used in engines designed for emissions at comparatively low piston speed. At higher piston speeds, the rapid filling of the cylinder is much more advantageous, and the combustion chamber design (which includes the piston crown) has a strong effect on the efficiency of combustion events. Fast-burning chambers are typically favored for high power output at high engine speeds. The burn rate of the fuel is also a factor to consider.

## How to Lift the Valves

If you are going to do any testing on a flow bench, you will need to have some method of opening (lifting) the valve. The valve should be fitted with test springs that are not so strong as to deflect any indicator or opening device. It is an easy process to apply, but most people purchase the opening devices and dial indicators so that they can get into testing sooner. It is worthwhile to note that a 1/4 x 20 UNC fastener has 20 threads per inch and one full turn will change the position 0.050". That doesn't mean you shouldn't use a dial indicator; it is just a suggested reference that is inexpensive and effective. The dial indicator is still needed for precision and repeatability.

Bill Jones, Precision Measurement Supply, Cal Spec, and many others offer opening devices for the valves in flow testing. Many are listed in the resource section in the back of this book.

It is very common to open the valve 0.050" at each test point, but an easy way to normalize the test points is to relate them to the valve diameter. If you use that method, and use 5% of the diameter for each point, the measured lift point will vary with the diameter of the valve. Such as 5% of a 2" diameter valve is 0.100" (0.05" x 2 = 0.100") and 5% of a 2.1" diameter valve is 0.105". Regardless of which method you choose, do the same thing in the same way so that the comparison of data is much easier to work with.

## What Makes a Quality Valve Job?

It is not my intention to attempt to show you how to do your own valve jobs. That would take another whole book. But I would like to give you some guidelines on what constitutes a quality valve job and perhaps encourage you to make some

# Flow Testing Intake & Exhaust Components

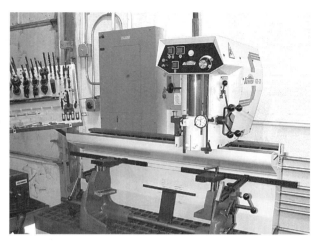

Here is a machine that is also used to cut valve seats and to install removable seat insert rings. It is very precise and can place the height of the seats within +/- 0.001" of all the others in a head. This is neat machinery.

 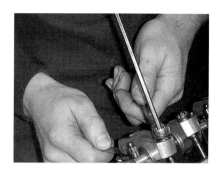

Setting valve lash is greatly taken for granted. There is a much more precise way to set the valves than using a "feeler" gauge. However, you can accomplish positive results with feeler gages if you practice. The text tells how to do it with accuracy.

Checking the valve seat run-out is an absolute necessity if you want to maintain quality control in your engine. Many folks scoff at this type measurement, but the lady making this measurement knows better and checks everything at least twice.

simple measurements of your own. Cylinder head work is difficult to do, and a lot of attention to detail to achieve the intended results is required.

The primary function of the valve job is to seal the valve, and even a poor valve job might still seal with today's high-load valve springs. However, there is more to a proper valve job than just sealing. Keeping the widths of seats uniform is another key issue as is the installed height of the valve tip from the seat. This is particularly important with hydraulic lifter applications or on non-adjustable rocker arms. The same considerations should be made for overhead camshaft (OHC) applications where the cam rides directly on the bucket that actuates the valve.

Remember these things are going to go clickety-

clack at very high rates, and there should be no compromises in the details if you are going to attain your goals.

The quality of the valve job is typically evaluated by surface finish and concentricity. The concentricity is defined as having a common center reference, and is essentially how much the seat is "out of round" with the centerline of the valve guide bore. If you ask your head shop what they use for a concentricity gauge or meter and they don't know what you are talking about, take your heads to another shop. Even automatic machines need to have the results checked on a regular basis. Modern valve springs will force the valve to seal even if the seat is not very concentric. The result is that the valve is bending and sliding into a sealing position because of the spring force. Eventually this banging of the seat and valve will cause wear and failure of the valve head or at minimum cause stability problems.

You do not need to know how to physically accomplish the valve job, but you do need to know how to verify the results and choose a shop that can do the best work.

***Three Angle Valve Job***—There is not a very realistic way to perform an effective valve job without using at least three separate angles. So, the expression "three-angle valve job" is really redundant. There are also some (unscrupulous or ignorant) shops that will provide a "cheap valve job" using a freshly ground seat angle only. That is not the correct procedure. A real valve job requires that there is a top cut, seat cut, and a bottom or narrowing cut. Although the specific angles might change, the previous descriptions would typically be a top cut of 15 degrees to 30 degrees (sometimes both). The most common seat cut angle would be 45 degrees or 46 degrees and the bottom or narrowing cut would be typically 60 degrees. In use today in many racing-only applications are seat

# Engine Airflow

Precision Measurement Supply makes a very convenient and precise tool to help you accurately set valve lash. It is the only way to get consistent settings for each valve. Once you try this, you will not rely on feeler gauges unless someone borrows your valve gapper. Don't loan it out.

These intake valves are for military aircraft. The valve on the left is from a WWII bomber's 2600 cubic inch 14-cylinder radial engine, and is 3.100" in diameter. The valve to the right is from a small four-cylinder engine for a UAV (unmanned aerial vehicle) and is only 0.965" diameter. Same job, different times and different sizes.

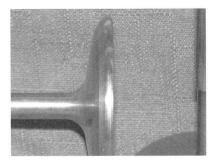

This titanium valve is larger than the normal valves made for a two-inch seat but flows much better because of the shape. Look closely at the shape and think airflow. This particular shape picked up the low- and mid-lift airflow about 10% over a conventional valve shape.

It is common to have a backcut on an intake valve that just intersects the seat angle. This backcut angle might be 30 to 37 degrees and not very wide. It should be flow tested for what works. Here the backside of a valve is getting a swirl polish/shaping treatment with a sanding roll and a hand-held drill in a vise. Go easy on these type mods until you see what works best.

angles of 50 degrees to 55 degrees, so the other cut angles would vary accordingly. Well-equipped shops can use a machine to even radius the area of approach and departure from the valve seat angle.

## Checking Valve Clearances

Since I mentioned the clickety-clack racket of the valvetrain as the valve lash, I should also mention there is a better way to set and check the clearances than using feeler gauges. The method would be to use a special dial indicator design just for setting the valve clearance. Of course, you can eliminate any of this checking by using hydraulic lifters, which don't depend on valve lash clearances, however, adjusting the valve lifter preload is necessary if maxiumum performance is desired.

Various test equipment has been available over the years to verify and keep close control over the valve lash (valve clearance) when setting valves. Precision Measurement Supply in San Antonio, Texas makes a nifty tool. If you check the valve clearance with one of these precision instruments and then check again using feeler gauges, you'll be surprised at how much variation there is in the readings.

## Valve Sizes and Shapes

The size and the shape of the valves have an effect on how the air flows past them. Changing the shape to improve the airflow requires some flow testing. There are not huge gains to be made here, but every little bit helps when you're hunting for any power advantage. The best way to find out what works or not is to try various shapes on the valves and test them.

Based on my testing experience, it is pretty common to back-cut the valves with an angle of something less than the valve face angle. It is common for a valve face-cut of 45 degrees to use a small back-cut of 30 degrees to 37 degrees, so that the ground area is about as wide as 80% of the face cut or narrower. That is just a guideline, and it depends on the original valve shape.

At one time, a "good" ratio for the intake valve diameter was about 52% of the cylinder bore. Today, that number has grown to a bit larger percentage for race engines that use one intake and one exhaust valve per cylinder. Multiple intake-valve cylinder heads have a distinct advantage over the single intake valve heads because of the area gain. Over the same time period, there has been a trend toward larger bore sizes. Since power is proportional to piston area, it was a logical conclusion to head in that direction.

Oddly enough, a very good shape for an intake valve is to have a valve diameter that is much larger than the seat diameter, so that the valve can be shaped to improve the exit losses of the valve and seat combination. A valve diameter of approximately 2.070" for a seat diameter of 1.97" sounds odd until you take a close look at the titanium valve shape (above lower left photo). That type design decreases the exit losses into the cylinder past the valve and seat. However that design also causes some other problems to occur. The valve described is typically heavier than a conventional intake valve but it is not uncommon for the shape described to improve the low lift flow at 10% or better with no loss at maximum lifts. Because of the shape of the described valve, it is

# Flow Testing Intake & Exhaust Components

Valve materials are many and a great deal depends on the intended use. Material selected for a million-mile diesel would not be the same as for an engine that only needed to run a 500-mile race.

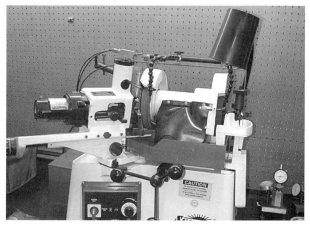

Valve grinding machines are important in a quality valve job. The grinding stone must be refaced often to stay true and must be changed if grinding titanium instead of other valve materials. The tool to the right of the machine is for verifying that the valve is straight and the ground surface is round. Both are critical in seeking quality work.

possible that there would be a problem with valve relief depths in the piston. This type valve design is just another compromise that might be worth considering for some applications. It has its limitations and it has its advantages.

## Valve Materials

There are literally hundreds of alloys that are used for valve material. It is normally best to rely on the suggestions of a quality manufacturer. Don't select the valve material based on whatever cheap deals you might find in the bargain basement. Many engine builders spend years establishing a good working relationship with such suppliers, so in the final selection process, it would generally be best to leave it to them.

Most standard production engines use inexpensive and common alloys such as chrome-nickel steel. The most common materials are alloy steels (particularly stainless types) and titanium. Depending on the application for the cylinder heads, one of the best all-around materials is stainless series alloy. The intake valves usually use 21-4N stainless alloy or something similar.

Exhaust valves are different story. Supercharged and turbocharged engines are a little tougher on exhaust valves, so it is common to use Inconel (also a stainless steel alloy) or 21-4NS material in those cases. Some engines also use stellite material (alloy of cobalt-chromium and iron), although it is more common for this very hard material to be used on the face (sealing surface) of the valve and on the tip of the stem of the valve.

*Titanium Valves*—Titanium valve materials are also varied and generally selected in order to make the valvetrain lighter in weight. Titanium is 40% to 45% lighter than steel and is very expensive. It is not uncommon for a set of titanium valves to cost more than $1,500 for a set of 16.

A potential problem of titanium exhaust valves is that they can generate alpha case material whenever the valve surface temperature rises over 1,200°F. The alpha case material causes surface cracks that are initially about 0.001" in length (depth) and will continue to develop until the valve fails. Big-budget race teams can afford to use and replace titanium valves on a regular basis, but lower budget teams may find it more cost effective to use another valve that can be regularly replaced. Many used titanium valves make it to the used parts market, so buyer beware.

In my opinion, two-piece valves should be avoided. If you have to use titanium, choose single-piece valves. It is also critical to have compatible material for the seats if using titanium valves. Beryllium alloys or beryllium-copper seat inserts are commonly used with titanium valves, but they are not the only materials available. You should follow the recommendations of the valve manufacturer to make sure the seat insert material is compatible.

Some stainless intake valve designs are supplied with hollow stems to decrease weight. Aircraft engine exhaust valves are sometimes hollow with an amount of sodium metal and sealed in place. At operating temperatures the sodium melts and becomes a liquid and helps transfer heat from the stem. Ford used some sodium cooled exhaust valves in some of their V8 maximum effort endurance engines years ago.

# Engine Airflow

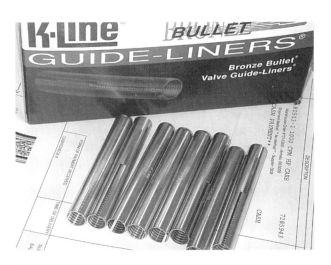

**Thin-wall valve guides such as these are normally inserted into a drilled and reamed hole then pressed in before final sizing. These actions are always done before a final valve job.**

Valve guides can be shaped in a lathe or mill before they are installed in a cylinder head. Careful selection of the material for the valve guides and the clearance of the valve stem in the valve guide are critical in quality work. Photo courtesy Endyn.

**Close up shows a bowl shape after cutting on a seat and guide machine. It also shows the thin-wall bronze material guide in place. Guides are made in all sorts of materials and sizes. They are typically sized with a hone to fit the stem of the valve with the proper clearances. Photo courtesy Nat's Engines.**

Some valve stems are typically chrome plated for wear resistance and some designs also use lash caps that fit on the end of the stems for improving the wear and the valvetrain geometry. Some titanium valves are hard surfaced using a nitride finish. The resultant titanium-nitride (TiN) surface is gold in color and is easy to recognize. The surface hardness of TiN is approximately 85 on the Rockwell C scale, which is very hard. However, almost all titanium valves use hardened lash caps to keep from wearing the tips of the valves.

## Valve Guides

The valve guides and the materials used for them can greatly have an effect on wear and stability of the valve during engine operation. The valve guide has to be loose enough to fit nicely when the engine is running at operational temperatures but tight enough to provide good stability and heat transfer and decrease oil paths into the combustion chamber. The fitment of valve stems to guides is somewhat of a closely guarded information item in some shops as is the material that they prefer to use. However, here are some bits of information that might help you make good decisions for your projects.

*Clearances*—The normal guide-to-stem clearance is from 0.0005" to 0.0015" and is greatly dependent upon the material and the application. Stock guide-to-stem clearance is typically 0.001" to 0.003" in production parts (cast iron guides and hard chrome-plated stems).

If your machinist knows what he's doing, he will offer advice on reaming vs. honing guides for precision valve stem-to-guide clearances. The overall stability of the valvetrain at high rpm depends on these kinds of details. The issue is not much of a problem on a pushrod motor with seat loads of 100 lbs. and valve lifts of 0.400". However if you plan on buzzing the engine to 8000+ rpm and have valve lifts of 0.650" to 0.800", then those extra details will pay off in good longevity and power. Imagine the problems encountered in Pro Stock engines with over one inch of lift at 9500 rpm, with a valve diameter of 2.5" or 2.6"! Can you say on the edge?

*Guide Materials*—The most common material for valve guides is cast iron. The advantage of cast iron is that it wears very well in stock configurations for hundreds of thousands of miles if the valvetrain geometry is very good.

Some replaceable guides in are made of Meehanite, a material that is an austempered flake ductile iron. The most common material upgrade in cast-iron cylinder heads is thin-wall inserts made of bronze or aluminum-bronze alloy. Do not be tempted by cheap brass guides.

There are many proprietary alloys that are used in these applications also. Some of the toughest alloys for guides and bushings are those made by Ampco Metal, such as their Ampco 18 and Ampco 45, which are nickel-aluminum-bronze alloys. They make a number of different alloys for various applications. The material is not inexpensive, but is worth some investigation to see if it would work for your application.

Most of the aftermarket cylinder head

# Flow Testing Intake & Exhaust Components

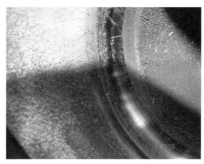

The bowl cutter is driven slowly in a machine that shapes the area and it saves time removing the metal. Courtesy Nat's Engines.

This close-up is after the bowl and seat area have been shaped by the cutters and is ready for the final valve job finish. Courtesy Nat's Engines.

Checking the seat run-out cannot be stressed enough. This type of concentricity tool is readily available at quality shops. This one is made by Sunnen and is very accurate. If it is not correct the seat must be redone to bring it into acceptable tolerances.

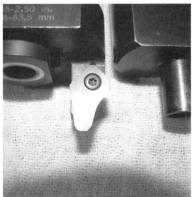

The seat cutting can also be done in a machine that uses replaceable form tools to establish where the final seat will be located. Look closely and you can see three angles and more to the shape cutting tool. Courtesy Nat's Engines.

There are many different form tools available for the machine that helps to form the critical area of the bowl and the seat for intake and exhaust valves. Courtesy Nat's Engines.

manufacturers and modifiers use guide material made from various silicon-bronze alloys. One very tough material is an alloy of copper-silicon-bronze and manganese. These specialty alloys are chosen based on experience using them in either long distance events, or with very high valve lifts and spring loads. The guide material must be compatible with the valve material (stem) that is going to run against it.

The guide and valve stem interface need to survive with only a tiny amount of oil providing local lubrication. Use the manufacturer's recommendations for clearances before you start getting too creative with making the clearances either less or more than that which is recommended. Your machinist will also be aware of not exposing too much of the alloy guide out into the exhaust port. Too much exposure out in the exhaust gases would encourage an inconvenient seizure of the exhaust valve.

## Valve Seats and Shapes

The seats can also be shaped to improve airflow, and since they work in conjunction with the shape of the valve, it is worth making sure they are complementary. Even at small amounts of lift, the shape of the valve and seat interface has an effect on the air flowing past it. The effects change with lift, and quite often a compromise must be used for the best results. There is some advantage to a radius-shaped seat, but that poses other problems, so a compromise in shapes is typically done.

One issue that is very important is the concentricity of the valve seat to the guide centerline. The concentricity should be held to no greater than 0.001" to 0.0015" run-out when using good guide material and clearances. Even in stock cast iron heads, the concentricity should be held to within 0.002" for best operation and longevity.

The seats concentricity is about 0.001" to 0.0015" on intake valve seats and not more than 0.0015" on the exhaust seats. That is a pretty tight tolerance, especially when you start working with very large intake valve diameters. The concentricity has to be within the fitted clearance of the valve stem to guide or the seat will wallow around and pound itself into submission or failure (whichever comes first). If the seat and valve and guide are all concentric to the centerline, the valve job will last much longer. And you can't win if you don't finish the race.

There are special machines that can cut the valve seat with a formed cutting tool that has multiple angles or even a full radius to and from the seat angle. These machines are so amazing that the valve job can be done and all the seats are cut easily to within 0.005" (or closer) of all the rest in the head.

# Engine Airflow

Never take anything for granted. Measuring valves for straightness and overall condition after grinding or even if they are new is mandatory if you seek high quality in the valve job. This is just as important as checking the seat's run-out.

Final seats are sometimes ground with a seat grinder and it is used to just touch the formed seat so it is a finer finished surface.

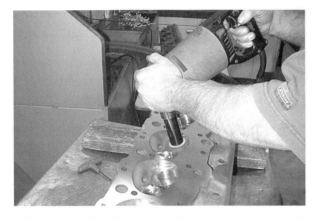

The ground seat and a top cut are shown here in this close-up. The seats are checked for run out and concentricity after this work. Seats are about 0.040" wide at this point. Angles here are 45 degrees for the seat. The top cut is 35 degrees, but you are looking at a total of six angles here.

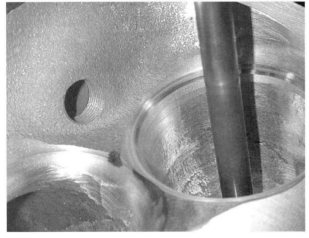

This cylinder head is a high-port 23-degree small-block Chevy, and it is one of the 867 Pontiac castings that were quite the rage in NASCAR years ago. It has been ported by Mike Chapman. The seat grinder in front of the head is a Kwik-Way SGF style. Hard to use but very accurate. This head tested to 310 cfm at 28 in.H2O at 0.700" valve lift. It should make good power.

They are worth the added expense over doing it entirely by hand.

There is an old engine builder's saying that states: "Long race = wide seats, short race = narrow seats," and although there is some truth to that, there are other ways to look at the problem. That is something to consider on air-cooled (oil-cooled) cylinder heads. If a combustion chamber has adequate cooling passages around it, and the valvetrain is stable at high rpm, the seat's widths are not as critical as they once were. It is now common on drag racing valve jobs to have a seat width as narrow as 0.015" to 0.020" on intake valves and 0.030" to 0.040" on exhaust valves. A much more common number on widths for intake valves seats is 0.040" and 0.060" to 0.080" on exhaust seat widths. As you can see here, there are always other approaches to reach the same goals.

*Port Volume*—This is a reference to how much volume (normally in cubic centimeters) that a port holds with a valve on the seat up to the cylinder head intake manifold flange. The port volume doesn't really relate to very much of significance, but is a commonly used reference. As an example, if you were talking about 23-degree Chevy heads and said the intake port held 200cc, that doesn't tell you very much about the shape of the valve or the bowl or the port. It simply states how much liquid the port was measured to hold. It certainly does not relate to average velocity nor would it compare to another type of cylinder head with a similar volume. So, where did the reference come from? It is mostly used as a reference by NHRA technical inspectors to distinguish between cylinder heads for the Stock and Super Stock drag racing classes.

The performance aftermarket has also adopted port volume numbers as a reference. One thing for certain is that the port and runner of the manifold are a continuation of the same port with only the gasket material separating them. So perhaps the volume of the whole port should be considered at some time or another. Maybe when thinking about the tuning characteristics of the intake and exhaust systems is a better place to discuss this issue.

# Flow Testing Intake & Exhaust Components

Properly matched valvetrain components are an absolute necessity in any engine but particularly in a racing engine. Most of the name manufacturers want you to be successful with their components so they put in a lot of effort to make sure the components work as a well coordinated package.

Valve springs are the components that must be watched carefully in routine maintenance. Initial selection is critical and should be left to experts. The small springs on the left are for Junior Dragster engines and the triple spring packs on the right are for very high rpm engines. Springs are super critical and necessary valvetrain parts.

Testing springs becomes a normal task whether in the shop or at the racetrack. Attention to valvetrain maintenance cannot be compromised if you want your engine to live long. It is strongly suggested that you verify the springs even if they are new. Better safe than sorry.

## Valve Springs

The correct selection and installation of valve springs is very critical for the engine and its components. However it is one of the most abused areas of the valvetrain. The incorrect preparation and installation of the valve springs is so common that it baffles the imagination. The other most common mistake in assembly is having an insufficient clearance between the valves and the pistons. If the valve springs are not correctly selected and installed for the application, the valvetrain will separate (float) at some rpm and engine component destruction will be the result. That makes ricochet power.

Spring loads that are used in engines are all over the map and depend on the application. OHC (overhead cam) applications use less spring load than the pushrod operated valvetrains. The spring loads on pushrod engines vary from about 80 to 90 lbs. at installed height to more than 250 to 300 lbs. at installed height on racing engines. Matching the spring load capacity to the cam is a critical step in maximizing the performance potential of the engine. Spring loads on the seat are only part of the equation.

***Desmodromic Systems***—Unless you are fortunate enough to be working with one of the engine designs that utilize a *desmodromic system*, you will have to deal with valve springs. A desmodromic valvetrain does not use valve springs and instead uses a mechanical process to open and close the valve(s). The desmodromic valve actuation process

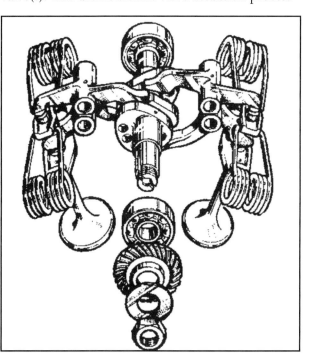

Desmodromic valvetrains have been around for a very long time and are impressive mechanical devices that use no valve springs per se. They only use some mousetrap springs to keep the overall lash contained.

Shape of a small-block Chevy combustion chamber with a note from the legendary Joe Mondello. Some day this kind of stuff should be in a museum.

dates back to 1907 and has been used on many historically significant engines. Mercedes Benz, Auto Union and OSCA engines (Maserati Brothers) were early examples. Ducati motorcycle engines have used the setup since 1956.

The valve springs and parts (retainers, keepers, shims, locators) are generally recommended by the camshaft manufacturer. The approach of "mixing and matching" of valvetrain components is almost always a sure recipe for disaster. All these parts must be complimentary or interrupting harmonics and a clash of frequencies will destroy the components long before they can make horsepower. The term "ricochet horsepower" generally describes the self-destruction of an engine. Stay away from the trick of the week.

*Pneumatic Valve Springs*—Formula 1 racing engines typically use pneumatic valve springs and an enclosed canister that is charged with a specific pressure of nitrogen, because it is not affected as much by changes in temperature. These pneumatic valve springs would solve lots of problems if they were allowed in U.S. racing, but that is not likely to happen anytime soon. Here again it is not technology that is the problem, but politics within the organizations.

Thank goodness testing cylinder heads on a flow bench does not require anything but checking springs. For flow testing, all you need is a valve light enough so it will open easily, yet stiff enough so the valve won't be pulled off the seat.

One thing to remember about valve springs is the negative effect blower boost has on intake valve springs. If you have intake valves that are 2.25" in diameter, and they have 0.341" diameter stems, the area of the valve is 3.9761 square inches minus the stem's area of 0.09133 square inches, which yields a geometric area of 3.8848 square inches. If the blower boost is 15 psi, that is a force of 15 x 3.8848 = 58.272 lbs. trying to push the valve off the seat. That takes away 58 lbs. from the normal spring force on the seat. Knowledgeable gearheads consider this when setting up the cylinder heads so that maximum performance is not hindered from premature valve float. That problem becomes more serious on engines that use two or more intake valves, because often there is a problem getting

This aircraft engine cylinder head has been cut from the cylinder (no head gaskets) and you can see it is an open chamber design of low compression ratio (6:1). The bore size is 5.250" and the intake valve size is 2.250". Normal max rpm is only 2100.

enough spring loads in short spaces like on OHC engines running at 25 or 30 psi, as many turbos love to do. The valve area issue is something you should always be aware of. Don't just look the other way and keep putting larger and larger turbos on and not gaining any power. It just might be the spring thing that is working against you. Do the math and make more power.

## Combustion Chamber Shapes and Sizes

Combustion chambers come in all shapes and sizes, which are generally dictated by the size and location of the valves. There are many engine designs that use combustion chambers located in the piston head, and the cylinder head is flat. The volume of the combustion chamber and its shape provide whether it is a fast- or slow-burn type. Tightly shaped combustion chambers are normally faster to burn than the open or lazy burn types. The configuration of the piston and the combustion chamber and how close the piston comes to the cylinder head in dynamic operation establishes how much spark advance is necessary to make maximum power. Normally, if a chamber is a fast-burn type, and if the engine is very efficient, less ignition advance is preferred.

Of course the size of the bore and the location of the spark plugs are also a factor. It is much better to have a design that encourages airflow (particularly around the intake valve or valves) than to have a higher compression ratio at the cost of improved airflow.

*Chamber Design*—The design of the combustion chamber can somewhat dictate how detonation resistant the engine might be (even at

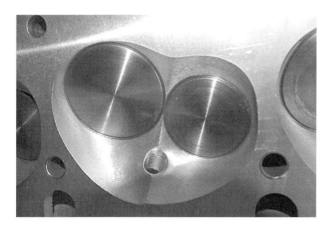

This special combustion chamber is for a big-block Chevy. Note the very gentle slopes to the sides of the chamber and the small size. This will allow a decent compression ratio without any dome on the piston at all.

This small-block Ford combustion chamber is small and well-shaped with no sharp places. The head is to be used on Mustang Sally, which is a Bonneville project that is only 292 cubic inches with a very high static compression ratio.

These combustion chambers have the intake wall pulled farther back, which helps flow as the intake valve opens.

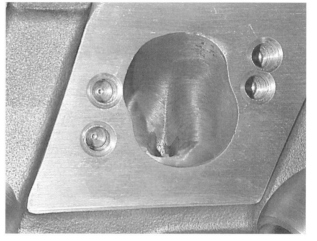

This exhaust port has been raised for this small-block Ford setup. It might look strange, but it also uses some exhaust header adapters external to the head. It does flow well. Head is by TFS.

high compression ratios). The geometric design of the combustion chamber can help encourage a fast and more complete burn of the fuel available. If the chamber shape is not conducive to a smooth advance of the burning mixture's flame front, it might encourage detonation or abnormal combustion even at low compression ratios.

A well-designed combustion chamber will have no sharp edges with the spark plug located in a central position. It is better to start the spark that starts the fire in a somewhat quiescent position (not prone to foul with fuel), and in a warm spot as well. The ideal chamber will be somewhat shallow (not a large volume), and encourage rapid burning movement in order to capture as much heat as possible to push on the piston without much heat loss. More efficient combustion chambers require less advance for the ignition spark. If an engine can make good power at 30 degrees BTDC, it is better than attempting to crowd 35 degrees of advance in it. The intended target is that after the ignition point and combustion begins the location of peak pressure after TDC would be something between 12 and 20 degrees ATDC with around 15 degrees ATDC being a very common result. The location of peak cylinder pressure is more important than just the maximum pressure, because it generates more area under the curve, which provides more average "push" on the piston.

## Intake vs. Exhaust Flow

It is very common for the intake port airflow capacity to be greater than the exhaust port airflow capacity. However, that is not cast in stone. What is the net result of having a higher flow rating on the exhaust over that of the intake? Well, it will require that the exhaust valve timing be somewhat different or the engine will over-scavenge the cylinder, reducing power output.

How much is enough? How much is too much? Those are somewhat trick questions, but I will attempt to explain in context with understandable answers.

# Engine Airflow

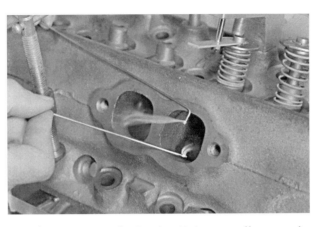

This shows there may be some surprises in airflow when you investigate it. The flow on this exhaust port is actually flowing at least two directions at once. Look at the photo closely and think about what you see.

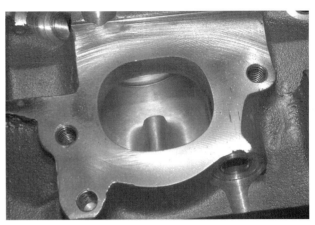

This is a highly modified BBC exhaust port. If you look closely, you will see that there is a sleeve installed on the right side of the port so a head bolt can be used without interference. This effort was required in order to keep the flow confined to the exhaust port as cast by the manufacturer.

*Exhaust-to-Intake Ratio*—It is generally enough exhaust-to-intake ratio (flow reference) to have the exhaust side flow at least 60% of what the intake flows. There might be small gains up to 70% on the exhaust side, but almost all very well rated cylinder heads have a flow ratio on the exhaust of 60% to 75% of the airflow of the intake side. Those numbers are with an exhaust pipe section attached (normally about 8" to 10" long is adequate). At the time of this writing, most successful drag racing Pro Stock cylinder heads have exhaust airflow to intake airflow ratio of about 60% to 65%.

Many novice cylinder head modifiers like to work on the exhaust ports because they are easier to get at than the intake ports. Another way to look at very high exhaust to intake ratio flow numbers is when the exhaust is very high, it might be an indication of just how weak or poorly developed the intake port is. The late circle track racing legend Smokey Yunick once stated that small-block Chevy heads used for 355 cid engines needed at least 250 cfm on the intake and 225 cfm on the exhaust, is a whopping 90% exhaust-to-intake ratio (that was targeted to make endurance racing horsepower at 8000 rpm). Today those numbers are not very common relative to the flow and power levels of similar racing engines so there must be more to the story.

## Establishing Cylinder Head Characteristics

Yes, there are cylinder head designs that have the exhaust ratio at 85% to 100+% of what the intake flows, but they require special camshafts or low ratio rocker arms to "calm down" the low lift flow on the exhaust side. In order to establish the characteristics of a cylinder head on the flow bench, one test can provide some valuable information before selecting a complimentary camshaft.

Place the lift of the intake off the seat about 0.100" and flow the intake in the normal direction. Record the results. Then reverse the flow on the bench so that you are flow-testing the valve and port in reverse (also at the 0.100" lift). Record the results. If the results show that the reverse direction flow test is higher in value than that of the forward test, you will have a potential problem of reversion with the combination. Do the test on both intake and exhaust sides of the cylinder head. If the engine is prone to reversion, the camshaft manufacturer needs to know so that he can suggest a solution for that particular combination. It is good to know at what lift the flow bias might be equal in either forward or reverse directions. This seems to be an issue with hemispherical or pent-roof type combustion chambers, because the exhaust ports are often on centerlines with their intake port counterparts. That causes the overlap event (the transition time when both intake and exhaust valves are off their seats) to allow mixture to ride right on through the chamber and out into the exhaust stream. This phenomenon is sometimes called "over-scavenging" and can seriously affect your horsepower numbers. It also wreaks havoc on emissions and fuel mileage.

A simple explanation here is that if the exhaust system is over-scavenging, the engine there is unused air and fuel that is tossed away without using it efficiently to produce power. Those engines would typically have a very poor BSFC number and a high BSAC number. After you have worked long and hard at getting the engine to breathe in more air and fuel there is little sense in throwing it away in an overly aggressive exhaust port and system.

*Cast Iron or Aluminum Heads?*—In Detroit-based engine designs, the cast iron material has been king from the beginning. Why? It is cheap

## Flow Testing Intake & Exhaust Components

Steve Garvey can weld all things aluminum. He is welding up the combustion chambers and ports of a cylinder head. Photo courtesy Endyn.

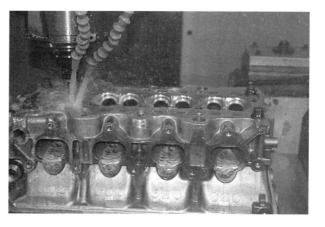

Use of CNC machines is a time-saver these days for shops that can afford them. This shows a previously welded head getting reshaped with the high speed cutting of aluminum. Photo courtesy Endyn.

After welding, the head is ready to go to the CNC machine for shaping of combustion chambers and ports. Photo courtesy Endyn.

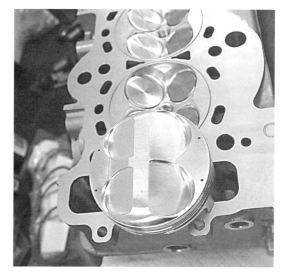

One of the advantages of CNC equipment is that the piston (made on a CNC machine) and the head can be precisely matched so compression and breathing are both maximized. Photo courtesy Endyn.

(relatively) and easier to cast in mass quantity.

However, for racing cylinder heads it certainly has some serious disadvantages. Namely, if you break one, they are very difficult to repair. Not impossible perhaps, but difficult (and expensive) to repair. There are many very good cast-iron cylinder head designs that are produced in the aftermarket. They are normally produced in much lower numbers than their Detroit counterparts and as a result generally producing more "dense" (less porous) castings.

Cast-iron cylinder heads are very good in that they are generally more rigid than aluminum alloys and provide somewhat of a strengthening "girdle" for the top of the block. That is not to say that there are not problems with using cast-iron material as the best choice for cylinder heads. NHRA Super Stock rules are pretty well attuned to the OEM cast-iron cylinder heads. Current rules allow porting and repositioning of the ports but afterwards the volume of the ports (in cubic centimeters) has to be equal to the listing that NHRA technical services provides. Inadvertently the ruling body (NHRA) created a real plus for the engine by requiring that the volume be the same as the original specifications because this alone increases the local velocity. That is one of the many reasons why NHRA Super Stock category vehicles are so impressive with their performances (these engines have to stick to the original compression ratio).

Even Detroit and major OEM vehicle manufacturers have applied new foundry techniques so that aluminum castings today are better than ever before. The aftermarket has almost exclusively chosen to manufacture their cylinder heads of various aluminum alloys. In general, the current crops of aluminum alloy cylinder heads are much lighter than any equal cast iron cylinder head.

At the time of this writing, there are more cylinder head manufacturers and selections available than ever before in motorsports history. You can get spread ports, high ports, small ports, huge ports and many, many variants in between.

# Engine Airflow

These are but a few of the tools used to reshape ports and combustion chambers.

To shape ports, you'll need a lot of time and a wide selection of "tootsie rolls," or cartridge rolls in various grits.

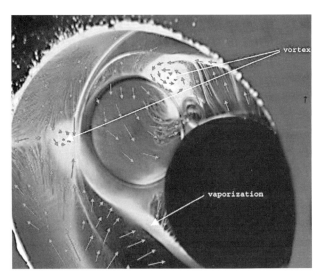

Wet flow in the combustion chamber can be confusing to understand. The amount of flow and counter flow in this space has all sorts of turbulence (vortices) that can be changed by altering valve seat angles, and sharp and rolled places in the combustion chamber. Photo courtesy Darin Morgan.

The dotted lines show where the port was originally located. As you can see, it has been moved considerably by welding and reshaping. This head is for a 500+ cubic inch street engine.

There is no single best answer; some compromises must be made and the skill is in selecting the stack of compromises that works in your favor and application.

The aluminum alloy cylinder heads are not without some problems of their own. The valve seat inserts are more prone to move around with heat cycles. Over time with exposure to high heat, they are more prone to change the material hardness with these cycles. However, the cylinder heads are much more convenient to weld up and repair or move and reinforce things than are their cast iron counterparts. Most modern alloy cylinder heads have excellent longevity and provide superior service if treated correctly. As with many things, abuse of components will not pay dividends on your engine. One of the saving graces of using aluminum alloy for cylinder heads is that it helps to carry away combustion heat very effectively so that you can generally run a bit higher static compression ratio than with the Detroit wonder metal (cast iron).

## Head Porting Tools

The right tool for the job makes the process much more effective and easy to accomplish the job as well. Don't think that the first thing you need to pick up is a grinder. That thought will cost you way more than it is worth.

Some of the things that you will need to have if you want to port your heads:
- Layout tools
- Grinders: Electric and Air
- Arbors and collets
- Cutters: Carbide and tool steel (use the right ones for the material to be cut)
- Tootsie rolls (abrasive sanding rolls with appropriate grits for finish required)
- Grinding stones: Shapes and materials more effective for cast iron
- Flappers and abrasive cloth
- Gloves
- Eye and ear protection
- Facemask/breather filter
- Sanctioning body rule book

If you think of the grinder and cutting tools as providing a way to shape the port instead of just cutting the material, you will become more adept at gradual and careful material removal. So that gentle and delicate shaping is the target instead of just "making the chips fly." Always remember that shape is more important than size and your ports will improve as a result.

## Wet-Flow Testing

Because airflow normally carries suspended fuel in droplet form with the air into the cylinder, if there are sudden changes of direction, the fuel might have a tendency to drop out of the flow path and splatter about the walls or cling to a surface. *Wet-flow* testing will help to identify those potential problem areas so that a fix can be done. Changes in angles or the like have influences on the wetted flow particularly along the walls, roof, and floor of the intake ports.

## A HEAD PORTING EXAMPLE

When Rich Albright at Midwest Cylinder Heads decided to make a few changes to improve the flow on some cast-iron 455 Oldsmobile heads he was working on he just started in and when he finished he had moved the intakes up 0.625" and the exhaust ports were moved up as well. The bottom side of the intake ports was filled with epoxy. The combustion chambers were completely redone too. The heads were welded with cast iron rod in a special oven set up. Final flow was up from about 200 cfm to 270 cfm. The head now uses a 2.05" intake valve and the 270 cfm happens at 0.700" lift while the exhaust flows 190 cfm (both at 28 in.H$_2$O water test pressure). Imagination with skill produces awesome results and should get some Super Stockers attention at least.

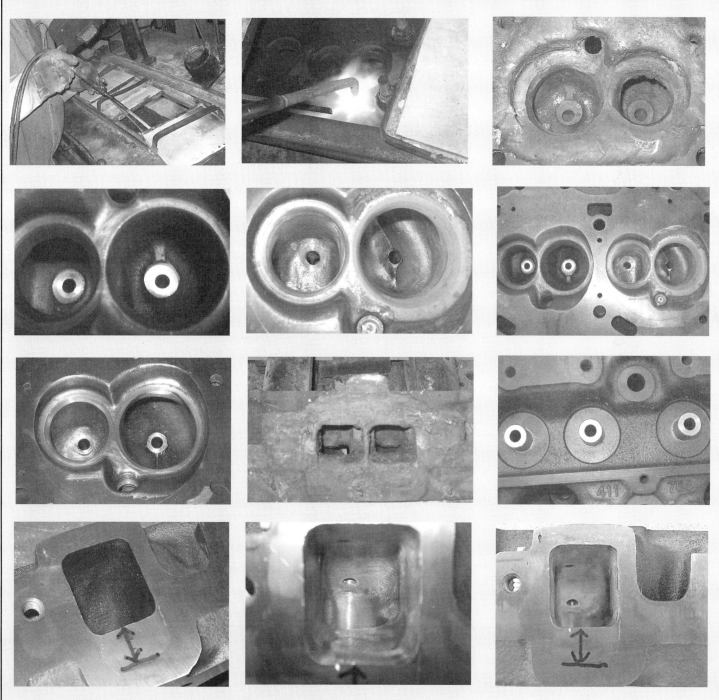

*continued on next page*

# Engine Airflow

*Head Porting, continued*

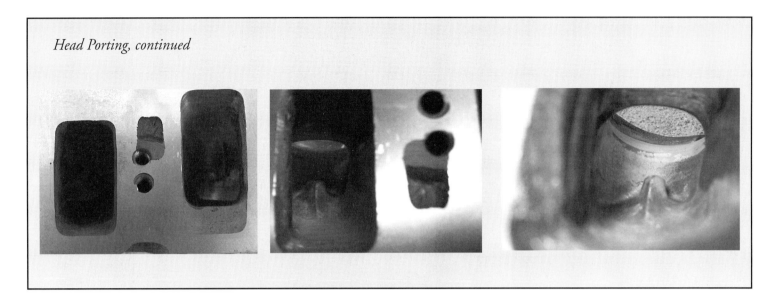

This is a fabricated plenum on a standard although reworked set of runners for a Honda four-cylinder engine. The design was found through careful flow testing. Photo courtesy Endyn.

The components that Joe Mondello and Lloyd Creek have developed to attach to flow benches can provide good information on wet-flow characteristics; it is certainly worthwhile to consider it on maximum performance engines. The wet-flow attachments can be affixed to any flow bench.

***Types of Intake Flow***—There are two types of flow in the intake port. One is called *spray flow* and is mostly down the centrally located portion of the airflow path, and the other is *wall flow*, which is mostly along the mechanical confines of the port. The dye used in the liquid with the wet-flow attachments will show where the liquid collects or puddles, both during and after a flow test. Slight changes can be tested to move or better disperse those collection areas and the fuel will become more burnable in order to more effectively feed the combustion process. The liquid film that forms on the port walls during transient cycles (varying engine rpm rapidly) can be more effectively controlled if the average airflow is at a high velocity, because the fast moving air and fuel helps to reduce the amount of liquid fuel on the walls and floor of the port.

## Flow Testing Intake Manifolds

Testing intake manifolds can be very frustrating. If the manifold is a single four-barrel common plenum type, the difficulty is in getting all the runners to flow well and be complimentary at the same time. Most manifolds have design criteria that sometimes conflict with performance. The manifold has to easily fit an engine configuration with minimum machine work, if any. On top of these compromises, it also has to live under a closed hood.

Sometimes the castings are designed for fit rather than for airflow. It is not uncommon for an intake manifold to reduce airflow by 10–15% or more when installed on the cylinder heads. But careful modifications to the manifold can reduce that airflow loss to about 5%. It is not easy, but it can be done. Proper selection of a manifold for the application is a good start.

But do not get lulled into believing that all you need to do is plop a manifold in place and expect the best results. Manifold manufacturers are forced

**This small-block Ford tunnel ram manifold has been reworked on the inside for better flow characteristics. The target is to reduce the loss across the manifold regardless of the configuration.**

# Flow Testing Intake & Exhaust Components

This small-block Ford intake port has a pretty big drop over the short-side radius into the valve seat and has some problems with wet flow. This intake port will be used with the tunnel ram manifold shown before. The head is obviously inverted in this photo.

This is a home built Bevers manifold for a Ford engine. It takes time and patience and it can be done if you have the skills. Photo courtesy Cathy Bevers.

Plenum and runner entry area of this manifold is welded and reshaped. The result was an impressive increase in flow numbers. Photo courtesy Meaux Racing Heads.

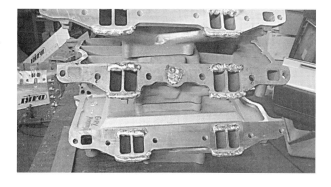

Welding up aluminum intake manifolds before working on the ports is a necessity on some applications such as this stack of units for a Mopar. Photo courtesy Meaux Racing Heads.

to make compromises to meet the demands of an easy-fit mass market. Some of the fabricated pieces are often designed for cost-effective mass production rather than for best airflow numbers. So as the engine builder or tuner, you need to squeeze out every bit of efficiency for your application. Checking with the fabricator for flow numbers and details is a good start.

Regardless of the manifold type, it is better to have the manifold mounted to a cylinder head for evaluation and testing than it is to toss the manifold on the flow bench just by itself to gather flow numbers. It is important to tape off any of the runners that are not being tested so that the flow path is properly simulated for any given cylinder.

When testing intake manifolds, you should also mount the carburetor or throttle body and evaluate it as a system and at various throttle openings. If you are attempting to improve the design of an electronic fuel injection manifold, you need to consider where and how to mount the throttle body or bodies. Did the OEM designer have maximum power in mind or was he trying to just get it all under the hood and clear the power steering pump and alternator? The location of the throttle body or the carburetor(s) can also have an effect on air distribution to individual cylinders. (See Chapter 10 for testing carbs, throttle bodies and EFI.)

Ideally, one would like to have all the cylinders getting equal amounts of airflow. However does the cylinder head have equal flow in all the intake ports? It is not uncommon for some cylinder heads to vary by 10% or more even after being carefully reworked. It is a real challenge to get the variations from port to port down to 3%, which is a very good goal to achieve.

*Manifold Modifications*—In general, the bottom of the plenum area of a common plenum manifold should be a radius shape that joins the bottom of the port entries. At the time of this writing, most common plenum manifolds have either flat or peaked bottoms in the plenums. The shape of the plenum will change when using a restrictor plate as required in some NASCAR categories. Restrictor plate type racing rules are used in everything from go-karts to limited Formula aircraft engines. Finding the best solutions for those applications requires a combination of testing on a flow bench, dyno and on the track in racing conditions.

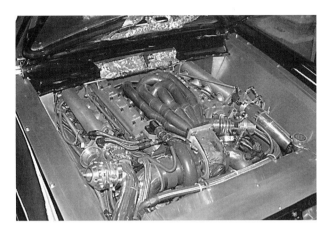

Turbocharged Ford Indy engine in the engine bay of this early Falcon is a tight fit. The planning, design and fabrication were done by Kenny Thompson. Photo courtesy Tom Monroe.

Kenny Thompson also did the beautiful fabrication in these 180-degree exhaust systems for the Ford GT40 out of Dennis Oltoroff's shop. The top photos shows rare Gurney-Weslake cylinder heads. The design of the exhaust system was for enhancing the power curve of the engine. And the sound will make you smile. Photos courtesy Tom Monroe.

## Flow Testing Exhaust System Components

Exhaust system components can be tested to evaluate which has the least resistance to flow air. Engines perform well with very little backpressure. Backpressure is the resistance of an exhaust system or component to flow exhaust gases. The first 1% loss in power happens as a result of roughly 1 inch of mercury backpressure, which as discussed another chapter, is equal to only 13.6 in.$H_2O$. Most OEM and some aftermarket exhaust systems have backpressure numbers that are 10 psi (276.9 in.$H_2O$) or more, which will substantially reduce the power capability of the engine. However, in street engines, exhaust systems must be designed to satisfy noise and pollution restrictions and rarely operate at WOT. But a properly designed and sized exhaust system can still be quiet without restricting power too much.

*Mufflers*—Mufflers can be tested for their restriction to flow air, but that will not predict the resulting sound level. However, the sound level is not a direct function of the restriction of flowing air. My suggestion is to select the components that will make the most power without making unnecessary noise. Don't make the mistake of associating loud with power—there's no such thing as audible horsepower.

Even turbocharged engines can benefit from a properly tuned exhaust system, particularly before the exhaust is ducted into the exhaust turbine housing. It is a common misconception that supercharged or turbocharged engines do not need to address the intake or the exhaust manifold design. Forced induction engines respond just like their naturally aspirated counterparts to intake and exhaust system tuning. The only thing different is the amount of power they produce, and it is often at the same peak power rpm reference of a similar unblown engine.

## Intake and Exhaust System Tuning

It stands to reason that there might be some optimal lengths and diameters for the intake and exhaust systems for the engine. It would be so much the better if there were a simple solution, but as in all things relating to engines, there have to be some compromises.

The worst intake systems to tune are those that use individual runners. These types are also generally equipped with either a carburetor per cylinder or a throttle body per cylinder. The same worst case can be said for the exhaust design that uses individual pipes per cylinder, which are also referred to as "zoomies." Each design can have an effect on torque peaks and they need not be at the same place in engine rpm. Collected systems (whether on the intake or exhaust) are generally better for both increasing and smoothing the power curve.

Because the engine operates through a fairly

# Flow Testing Intake & Exhaust Components

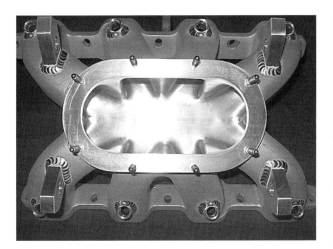

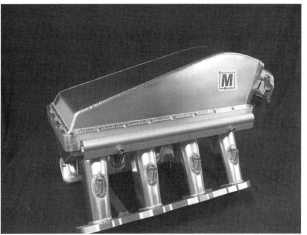

These shots are of a drastically reworked manifold for a small-block Ford with EFI. The top plenum was fabricated to work well with a throttle body mounted away from the plenum so that it was forward facing. Photo courtesy Marcella Manifolds.

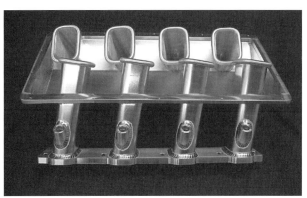

The plenum in profile does not expose what is inside, so it might look like just another manifold until you take the top off and see the runner lengths and the overall impeccable workmanship. For a Chevy LS7 engine. Photo courtesy Marcella Manifolds.

broad range of rpm, a simple single answer for the correct "tune" for an intake or exhaust system is very difficult at best.

Properly selected, intake and exhaust system components can enhance power production considerably. The selection of header lengths and diameters before collecting it all and then going through a muffler (and catalytic converter if required) before exiting the vehicle are all critical decisions. It sounds difficult doesn't it? You might use the following equations to select some headers to be helpful. Most computer engine simulation programs offer similar assistance. There are countless website resources, and all might get you in the same ballpark. Best performance is not related to cost only. It is also a function of what works best.

It would be nice if it were that simple. However, the subject is much more complex, so we will take a look at several approaches on how to calculate some lengths and diameters for intakes and exhausts and show some applications for each method.

*Freedom Systems*—The intake and exhaust systems can be designed to operate as single or multiple degrees of freedom tuned systems. A single degree of freedom system would be simply described as a spring and a mass so that the system would oscillate at a single specific frequency. Multiple degree of freedom systems are much more complex, but are commonly done for some applications.

Is it possible to tune the intake for one rpm and the exhaust system for another?

You bet. As a result, the torque curve of the engine might end up having multiple peaks. If those peaks are positioned just right, the careful tuning of the intake and the exhaust systems can broaden the range of the engine from peak torque rpm to peak power rpm.

*Intake Lengths and Diameters*—The intake system can be designed to operate as a Helmholtz resonator, a single pipe type system, or a collected system using a common plenum or any number of iterations of all those descriptions. Hermann von Helmholtz (1821–1894) first applied his resonator in 1863. The resonator was used to evaluate musical

**On individual runners on injectors the variation of length and diameters can alter the torque curve. This function holds true for all types of engines.**

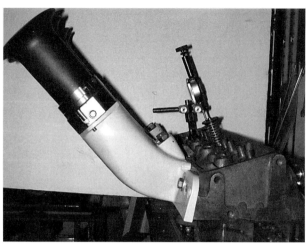

This Engler injector manifold is 14.5" long from the flange to the inlet radius. The cylinder head port is another 4.5".

notes, so his work in acoustics still applies today.

Each time the intake valve opens and closes, the impulse from that sudden valve movement sends a pulse of energy up and along the intake runner. The runner length includes the portion of the intake port that is in the cylinder head. A good way to think of the runner length is described as the length from the edge of the valve seat to the first rapid expansion for the runner (as into a plenum) or drastic cross-sectional change in area or to atmosphere. One thing that is common on all these tuning formulae is that the length of the port in the cylinder head is to be included in the length measurement. So essentially the length is measured from the valve seat closest to the intake side and continues to the first change of cross-sectional area (as into a plenum) or in the case of individual runners where it is first exposed to atmospheric air.

The length of the intake runners and the diameters of the runners have an effect on the "ram tuning" or "inertia supercharging" that occurs in the running engine. If the length and diameters are helpful and constructive, then the engine benefits from a little boost from a reflected wave of energy arriving at the intake valve just at the right moment to make sure the maximum amount of fuel and air are captured just as the inlet valve closes. The problem is that if you have a length and diameter that is helpful at some rpm it might not be helpful at others. In fact it might actually be detrimental at other rpm points. The intake system must be carefully considered so that the helpful boost that we want to happen does occur and good design would allow for not being overly harmful in other areas. It might be a good place to present some estimation of some cross-sectional targets for an engine:

$A_{im} = rpm \times S \times B^2 \div 190,000$
where $A_{im}$ = area of inlet (minimum) in square inches, rpm = revolutions per minute, S = stroke in inches, B = bore diameter in inches

$L = 84,000 \div rpm$
where L = length of intake runner in inches, rpm = engine revolutions per minute.

Note however that this doesn't give you any information on the diameter for the intake pipe. Likewise another quick and dirty method of finding out how long an intake system runner should be could be calculated by the following equation:

$L = 80,500 \div rpm$
where L = length of intake runner in inches, rpm = engine revolutions per minute. Or how about using $L = 82,500 \div rpm$?

Note that the spread on these previous three equations (at 6000 engine rpm) will result in a length of 14", 13.42", and 13.75" respectively. That is really pretty close (something like only 4% or so). Perhaps we can get closer agreement after some study and understand why.

So which one is correct? Which one would work best? There are several such equations that can be applied to both the intake and exhaust side of the engine, and each will give you an answer. None are perfect, but any would get you closer to helping the engine than just guessing.

Intake Length = $L = (cid \times 690) \div (d^2 \times T_{rpm})$
where L = length in inches, cid = engine displacement in cubic inches, d = port diameter, $T_{rpm}$ = Torque rpm

# Flow Testing Intake & Exhaust Components

One of the early OEM uses of "tuned" intake lengths was the 1957 Corvette with Rochester fuel injection. The runners inside the doghouse plenum were 7" long. Chrysler built the first production engine to exceed 1 hp/cubic inch with the optional engine for the 300B, which was a 354 hemi that made 355 hp.

Another unknown source uses:

$L = (90 \times V) \div N$
where L = intake length in inches, V = velocity of sound in feet per second, N = engine speed in rpm

Yet another anonymous offer is for individual runners such as fuel injection type setups:

$L = 93,500 \div rpm$
where L = inlet length in inches (includes the cylinder head port), rpm = engine speed in revolutions per minute

Dr. Helmuth W. Engleman (1920–1997) was a long-time professor at Ohio State University and spent most of his academic life devoted to studying engineering and internal combustion engines. He was one of the first to use cooking oil for bio-diesel projects after he retired. In his PhD. thesis written between 1952–1953, Engleman used the following equations for intake tuning:

$f = (C \div 2\pi) \times \sqrt{(S \div L \times V)}$
where f = frequency of resonance, C = velocity of sound in the gas, S = area of the throat or pipe in square inches, L = length of the throat or pipe in inches, V = volume of the cavity in cubic inches (cavity is the displacement of a cylinder at half stroke).

The ratio of the natural frequency of the cylinder and intake pipe is approximately two times the piston frequency. Chrysler used the following equation for years:

$L = 72 \times (C/N) +/- 3$
where L = intake length in inches, C = wave velocity (speed of sound) in feet per second, N = engine speed in rpm (where increase is desired)

Obviously the +/- 3 gives you some allowances in length variations. Over time, the +/- 3 term was dropped but is listed here to show that it was part of the original equation.

Seemingly with addressing more effect of camshaft timing, another length equation came about and it became the following form:

$L = 90 \times 1100 \div N$
where L = length in inches, 1,100 is the approximate speed of sound, N = engine rpm at tune point

Dr. Gordon Blair, Professor Emeritus, Queen's University of Belfast is a proponent of finite amplitude wave theory in the design of intake and exhaust systems. He is an acknowledged expert in the design and development of two-stroke engines. I first met him when I was in college as a neophyte engineer. He stated that the following equation was good for intake tuning:

$L = (1100 \times S) \div N$
where L = length in inches, S = number of crank degrees ABDC when the intake valve closes up to a maximum of 75 degrees. Anything greater than 75 degrees, use 75, N = engine rpm in the middle of the normal power range

*Helmholtz Resonator Manifolds*—The Helmholtz resonator types rely on the cylinder volume (normally at either full or half stroke) as part of the equation in calculating the runners and plenum(s) to be applied. However, there are many rumors about Helmholtz resonator models not working properly above about 3600 rpm to 3800 rpm. I personally don't believe them; I think such rumors were perpetuated by professional jealousy or confusion over the complexity of the calculations. There are many examples where some very effective Helmholtz tuned packages were more than effective at very high rpm, something like picking up over 30% in torque at design speeds. There are some SAE papers that support those higher revolutions numbers.

The real problems with the resonator-type manifold designs seem to be in matching the resonance properly and packaging the manifold to

# ENGINE AIRFLOW

This manifold was built by Wilson Manifolds for a carbureted set-up that also uses nitrous oxide. Everything has to fit in its place. Photo courtesy 10 Litre Performance.

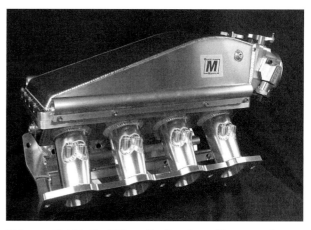

This manifold is for EFI application for a Mopar engine. After previous photos and explanations it makes you wonder what is inside doesn't it? Photo courtesy Marcella Manifolds.

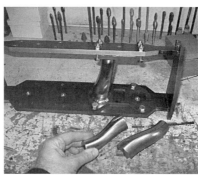

Roland of Virler and Bartlett fabricated an intake and exhaust (all stainless) system for a world-class race at Monza, Italy. The beautifully handcrafted piece helped to win the race. Photos courtesy of Virkler and Bartlett.

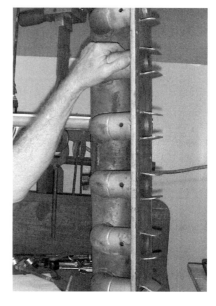

This is a common plenum intake manifold with runners that will end up on a twin diesel engine Bonneville streamliner (one manifold per engine). Yeah, this intake tuning stuff works on diesels too. Photo courtesy Lawrence Engines.

fit in spaces available. If the velocity of the runners is not done properly, the system will not function properly, and the ratio of velocity to volumes and lengths is what makes these designs work. If the wrong things are compromised or miscalculated, these designs won't work correctly. When the valve is closed, the column of air reacts as 1/4 wave organ pipe, and while the valve is open the Helmholtz tuning resonance and those complexities are in effect. In other words there are multiple actions with waves going on simultaneously.

A single degree of freedom calculation for the intake tuning can be done with the following equation:

$$N_p = 77 \times C \times \sqrt{(A_1 \div L_1 \times V_d) \times (CR - 1 \div CR + 1)}$$

where $N_p$ = tuning point in rpm, C = speed of local sound, $A_1$ = average area of runners and port in square inches, $L_1$ = length of port and runner in inches, $V_d$ = volume of cylinder at half stroke in cubic inches, CR = static compression ratio

Speed of sound = $49 \times \sqrt{T}$
where T = °F + 460

There are even some various tuning effects in consideration of the volume of the plenum and plenum volume can help out with the tune-up as well. The engine will respond favorably to a properly sized plenum and change its characteristics with a change in plenum volume. Often the tuner does not have much of a way to change the plenum volume (particularly on single four-barrel manifolds) but spacers can be added or deleted beneath the carburetor. On an EFI-type manifold, small changes can be made by adding or deleting spacers under the throttle body if the throttle body is attached directly to the plenum of the manifold.

*Plenum Volumes and Calculations*—The effect of the sizing of the plenum on the engine manifold design/tune is so that the plenum can help to remove some of the drastic pulses that the intake manifold experiences. Larger plenums dampen the pulses more than smaller plenums, but a plenum that is too large makes the manifold react as if the runners experience atmospheric pressure at the entry into the large plenum and the potential positive effects of the plenum are lost. It is not uncommon for a common plenum type manifold to improve the actual flow of the individual cylinders by up to 25% over individual runners only. That is a huge improvement for something so easy as sizing the plenum. In the case of the Helmholtz tuning models, the plenum volume should be sized so that the assistance the plenum provides should be at the rpm targeted for improvement.

The plenum volume is the volume of all the runners plus the plenum itself. The plenum impedance (ratio of flow and pressure) is calculated by:

$Z = (1 \div j) \times F \times V_p$

where $Z$ = plenum impedance, $j = \sqrt{-1}$, $F$ = frequency, $V_p$ = plenum volume in cubic inches. This is only shown so that you can have an appreciation of the calculations available. Note that the reference to the square root of a negative 1 is an imaginary number.

## Exhaust Lengths and Diameters

One could assume that the exhaust design must not be that important because most OEM designed exhaust manifolds that are cast iron generally are fairly simple and compact. However, OEM engineers are often more concerned with different issues, and one of them is weight, so smaller manifolds weigh less. The OEM folks also are not designing the package for the best configuration at all engine speeds and loads. They generally work within much more constrained rules than what we gearheads want to deal with.

The exhaust system will typically produce less performance increase from tuning efforts than the intake systems show with equal tuning efforts. Any

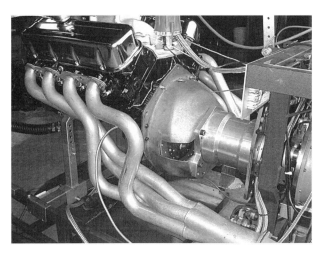

Exhaust headers become part of a given combination when testing on a dynamometer or race track. High on the list of importance is if they fit the race car. Testing is the way to resolve assumptions about what helps power.

improvements on the exhaust tuning mostly have an effect on only the pumping losses of the engine.

Many of the header manufacturers build their products to fit given chassis and engine combinations as the primary criteria and performance improvements are also in the mix, but not the only driving force. After all, the marketplace demands something that fits very easily and has some potential for performance improvements as well. So keep all that in mind when you go to work on your own calculations for lengths and diameters.

*Equal Length Primaries?*—Equal lengths are not as critical as you may think, as long as the configuration is complimentary to the engine and it fits in the space required. One of these issues of consideration is the sizing of the pipe itself, and more header manufacturers are offering multi-stepped pipe dimensions. As a result of using different material sizes and more welding, the cost of manufacturing these type of headers is more expensive than using simple straight-walled tubing. If you were testing on an engine dyno and found no difference in performance between a set of headers that cost less than a hundred dollars and a set that cost a thousand dollars, what would you think?

Some folks are even convinced that placing some little whirly-gig in each exhaust port or pipe will enhance the engine's performance. Some work had been done early on with a so-called Coanda designed semi-streamlined plug used in exhaust systems that was supposed to work, too. The designs were named for Romanian inventor Henri Coanda (1886–1972) who also flew the first jet-powered aircraft in 1910. There was also a time when Kadency-type tuning was quite the thing for two-stroke engines. There are many other names associated with this type of tuning over the years of engine development.

A flat-motor with Weber carbs, HEI ignition and collected headers is an odd find. It is not necessary to do like everyone else for a project to be successful.

One formula that has been used for header primary lengths is the following:

$L = (cid \times 1900) \div (d_e^2 \times rpm)$
where L = length of the exhaust pipe (includes port in cylinder head) in inches, cid = engine displacement in cubic inches, de = inside diameter of the exhaust pipe in inches, squared

A simple formula for header collector length is:

$L_c = (cid) \div (d_e^2 \times \pi)$
where $L_c$ = length of collector in inches, cid = engine displacement in cubic inches, $d_e$ = inside diameter of exhaust pipe in inches, $\pi$ = 3.1416. Note that this formula is for a collector length starting point and there is no reference to engine rpm

Another formula from Blair is:

$L_{ft} = (135 \times S) \div N$
where $L_{ft}$ = length of individual exhaust pipe in feet from the exhaust valve seat to the end of the pipe, S = number of crank degrees from exhaust valve opening to intake valve opening, N = engine rpm in the middle of the desired power range. If megaphones were used on each of the pipes the formula becomes $L = (150 \times S) \div N$

Yet another formula for length of exhaust pipes is:

$L = 205{,}000 \div N$
where L = length of the exhaust pipe (including cylinder head) in inches, N = engine rpm

Another exhaust pipe calculation, this one from British tuner, author and engineer A. Graham Bell:

$L = [850 \times (360 - Evo) \div rpm]^{-3}$
where L = header length in inches, Evo = exhaust valve opening point, rpm = engine speed revolutions per minute

Bell's exhaust pipe header diameter calculation is the following:

$D = [(Cyl \times 16.38) \div (L+3 \times 25)] \times 2.1$
where Cyl = cubic inch displacement of one cylinder of engine, L = length of exhaust pipe in inches

How about some inputs for collector measurements?

$C_d = 1.9 \times L$
where $C_d$ = collector diameter in inches, L = length of exhaust header pipe in inches

And if you want to estimate the collector length for headers you can continue using Bell's approach:

$C_l = 0.5 \times L$
where $C_l$ = collector length in inches, L = header pipe length in inches

As you can see, there are many approaches toward that point of simplicity where you can calculate, but it might not be an exact answer relative to the other solutions available to you. I might suggest however that you calculate using several of the methods listed and see how close they happen to be in their predictions for the length of inlet and exhaust systems. Opinions are sort of like the common ground for humans as we all have one or more. Which one is the answer that works best for you is up to you to decide. After you get a chance to work the math a bit I am willing to bet that you will begin to see some things surface that are common across the board of applications and that will lead you to make a sound decision. Better yet is if you have a chance to do some actual testing of which format works the best for you. Happy calculating and number shuffling!

# Chapter 10
# Flow Testing Carburetors, Fuel Injection & Manifolds

This organized leak is an important part of the airflow path on an engine. Flow testing the carburetor for airflow is just part of the answer. You should also know what the fuel supply side does in order to have an effective combination.

*It would be better for the true physics if there were no mathematicians on earth.* —Daniel Bernoulli

There are many ways to test carburetors and carburetor components on a flow bench, but making it easy also makes it more fun and quicker to learn about the carburetor itself. The carburetor can be tested for airflow and the effects of the metering circuits can be verified as well. The first objective in carburetor work should be to learn about the individual circuits and to learn how each circuit functions. If you don't do that, you will get miserably lost with what is going on in the magic mixer.

There are several ways to test carburetors on a flow bench and there is more to it than just measuring how much total air the carburetor will pass (that is the easiest part to test). The capacity of the flow bench is one of the limitations for testing carburetors. If the flow bench has insufficient capacity to test the whole carburetor at wide open throttle (WOT), the capacity of the flow bench is inadequate, so the unit will have to be tested either one throttle bore at a time or two throttle bores at a time. Either way, the carburetor unit needs to be adapted to the flow bench so that the testing can be done. It is just another little problem to be resolved so you can get back to chasing the tails of horsepower on the loose.

## Fuel Flow Data for Standard Holley Carburetor Float Bowl Components

To my knowledge, the test results in the chart on the next page have never been printed before this book was published. I'm providing them here so you can have better

> **Warning!**
> In order to flow test carburetors on a flow bench, the carburetor must be emptied of all liquid fuel and allowed to dry so that there is no liquid fuel or vapor to induct into the flow bench. Also the individual manufacturers should have safety instructions that apply to their particular designs of flow benches when testing carburetors or carburetor components.

data for planning your fuel system.

Liquid flow tests were done using a single Holley center pivot float bowl with clear cover (on carb body side) marked for float level drop and various standard needle and seat assemblies. The float level height was checked as a normal bowl would be verified. The top of the float was about 0.250" from the inside top of the bowl.

The fuel pump used was a single MagnaFuel (MP-4450) set at a system pressure of 25 psi to the regulator via a #10AN supply line. The regulator used was a single MagnaFuel 4 port design using only one outlet (#6 AN). One #8 AN fitting and line out of the bottom of the modified float bowl to a gate valve controlling flow out of the vessel (float bowl). The gate valve was opened and closed to simulate the use of fuel, as an engine would have different demands.

Flow measurement was accomplished with a highly

# Engine Airflow

Turbocharger compressor housings can be tested and quantified on a flow bench, which adds to the overall quality control on a project using that type of equipment. Intercoolers can also be tested for flow losses.

A carburetor can be tested on the flow bench using the complete fuel circuit (signal only with a dry carburetor) by using fasteners and spacers so the booster signal can be checked through the metering body and air bleeds. The tube goes to an external U tube manometer where the main circuit can be measured while on a flow bench.

### Holley Carb Fuel Flow Data

**Regulated Pressure @ 4 psi**

| Fluid Level Drop | Flow (gpm) | Flow lb/hr (gasoline)* |
|---|---|---|
| 0.100" | 0.048 | 17.86 |
| 0.200" | 0.198 | 66.96 |
| 0.300" | 0.317 | 117.92 |
| 0.400" | 0.417 | 155.12 |
| Float on bottom | 0.528 | 196.42 |

**Regulated Pressure @ 6 psi**

| Fluid Level Drop | Flow (gpm) | Flow lb/hr (gasoline)* |
|---|---|---|
| 0.100" | 0.158 | 58.78 |
| 0.200" | 0.256 | 95.23 |
| 0.300" | 0.422 | 156.98 |
| 0.400" | 0.528 | 196.42 |
| Float on bottom | 0.687 | 255.50 |

**Regulated Pressure @ 8 psi**

| Fluid Level Drop | Flow (gpm) | Flow lb/hr (gasoline)* |
|---|---|---|
| 0.100" | 0.132 | 49.10 |
| 0.200" | 0.375 | 139.50 |
| 0.300" | 0.441 | 164.05 |
| 0.400"0 | .499 | 185.63 |
| Float on bottom | 0.763 | 283.84 |

*0.743 SG (6.2 lb/gal)

accurate positive displacement flow meter that displayed results via a liquid crystal display (LCD). NIST traced accuracy was certified to be linear within +/-.2% within the range of the test flows shown.

The needle and seat combination for the tests in the chart at right was a standard Holley 0.110" diameter with Viton tip. All tests were run with the float level set to the visual reference of the bottom of the sight gauge (per normal field operation of carburetor users) and at 0.250" from the top of the float bowl casting.

Foaming occurred at any greater liquid level drop than 0.200". Severe foaming occurred at liquid level drops greater than 0.300". Note the flow results at 8 psi. Most of the disparities in flow versus tests done at 6 psi were probably as a direct result of the foaming that occurred.

In my opinion, the needle and seat combination should be as large as possible so that the float drop wouldn't be as much as in the test and would provide better fuel foaming control as well. There are available some needles and seats that are 0.125" to 0.130" Viton tipped units that work well for gasoline. There are also available large diameter metal-to-metal contact needles and seats that range from 0.130" to 0.150" for use with alcohols. The fuel pressure does not need to be set to a high number if the fuel system is capable of delivering enough fuel flow at lower fuel pressure settings.

One of the things that should be obvious from this series of tests is that the normal vibrations that are produced in and on an engine are probably easily transmitted to the fuel in the bowls or any device that is attached to the engine. If a pressure regulator is rigidly attached to the engine, it is certain that some vibration can be transmitted and will affect the regulator diaphragm/spring combination. Likewise, the carburetor should be isolated in some fashion from receiving the engine's various vibrations. Remember that vibration will produce heat and aeration in the fuel even before it gets into the passages of the carburetor that are supposed to accomplish the emulsion (mixing air into the liquid fuel) process.

# Flow Testing Carburetors, Fuel Injection & Manifolds

Don't assume that you have enough fuel supply just because a pressure gauge is showing some number. Flow and pressure are both factors on vehicle or dynamometer. There is more to fixing the problem than just setting the float level.

There really is a reason to use MaxJets or something like them in order to accurately select the flow capacity of the jets that your "carburnator" needs. Small increments of fuel flow change are important when tuning for maximum performance.

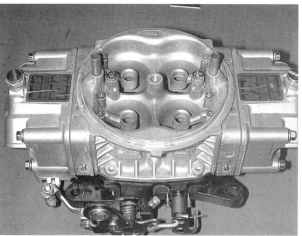

Carburetors pretty much look the same externally. On the dyno or the racetrack the evaluation takes on a completely different character. The reason these things are not cheap is because of all the R&D that goes into making them competitive.

Because of the specific engine excited vibration issue, some very skillful engine builders select various methods to dampen (isolate) the carburetor from the harmful vibration. The normal solution is through the use of a special neoprene material that doubles as a gasket between the carburetor and the adaptor or the manifold or both. The vibration dampening is accomplished by bolting the carburetor down only with enough force to provide a gasket seal and the carburetor essentially will be allowed to float on the dampening material. The tightening of the carburetor mounting fasteners actually makes the vibration problem worse, not better.

## Bench Tune Carburetors Using Simple Tests on a Flow Bench

For a long time, it was thought that there was not an easy way for the average racer to be able to size the jetting on a carburetor except on the dynamometer, the racetrack, or expensive wet-flow benches. The solution is particularly problematic if the carburetor was modified in some way or another. Come to find out, there is an easy way and the carburetor can be simply measured (on a regular dry airflow bench) and a jet size can be calculated that will be much closer than "in the ballpark" and substantially closer than guesswork.

*Carburetor and Calculator*—After several years of working away on the flow bech and building carburetors for a living, Steve Zicht, a carb modifier based in Virginia, has proven that a carburetor can be pre-tuned using his system of bench measurements and calculations. Steve literally wore out several hand-held calculators in the process of coming up with a way to take flow bench data and jet the carburetor very close to what the engine will want on the racetrack. Zicht has based his system on some very sound logic and has had a great deal of success in some real problematic applications. He has worked out the details so that the test time on the flow bench is as little as 30 minutes.

Although there is plenty of validity in fine-tuning and jetting on the racetrack, it would be nice if you could take a few simple measurements and jet the carburetor very close to what the engine will actually need to make best power before you leave the shop. The careful application of what you will read here could save you some trouble to say the least.

*How Carburetors Should Work*—The carburetor is not an overly complex device, but it has some real interesting subsystems that have to work properly in order to provide a decent mixture to the engine.

A simple description of the properly balanced carburetor is a metering device that meters the correct amount of fuel for the amount of air that it is ingesting. In order to do that, most carburetor

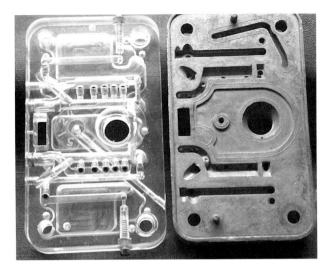

This plastic Holley metering block (left) was machined by my old friend, Bo Laws. It was done so I could use it for educational purposes. It is helpful to point out the many circuits that are functioning when a carburetor is working properly. Shown at right is the original metering block.

designs use a throttle valve (butterflies are not free), a low-speed circuit (idle and off idle), and a high-speed circuit (main jet and power circuit). Some have intermediate (part throttle) circuits. Most manufacturers "trim" the fuel flow circuits with "air bleeds" and orifice restrictions so that the proper ratio of air to fuel can be maintained as the throttle is moved from curb idle through wide open throttle (WOT). Some manufacturers also provide a series of bleeds (air leaks) into the main circuit that "emulsify" the fuel into small droplets like an aerosol spray. The common item on several of the circuits is the use of metered holes (orifices) to control either air, fuel, or both. The unfortunate condition of these items is that anyone with a drill set and a screwdriver thinks they can easily "fix" problems that previously did not exist.

***Test Procedure***—In most racing applications, the carburetor is most important during times at WOT. Drag racers can idle to the starting line, but once the green light flashes, it's hammer down on the throttle until the finish line. On a circle track, the throttle position dictates the speed of the race car. Part throttle is important, but not as important as the WOT call-up for maximum power. Maximum power is typically produced at around 12 to 12.5:1 air/fuel ratio (gasoline) and methanol usually runs best at about 5.5 to 6:1. Most of Zicht's customers are burning racing (or pump) gasoline, depending on the rules at the track, so he concentrated on solving the problem for gasoline first.

Steve wanted to make sure that the ratio of the airflow through the carburetor was correct for the fuel flow circuit. Many tests and thousands of data points proved to him that if he measured the signal at the main jet (measuring through the complete circuit), then all the air bleeds and miscellaneous holes could be accounted for. It was easy to measure the signals, but putting a procedure together that

made sense was the hard part. All his hard work and hours and errors have paid off. It works! Steve compared his data and decided to use the Max Jets as standards and referenced other jets if racers couldn't or wouldn't buy the Max Jets. He was impressed with the tight quality control and flow control when only making a small change (0.001" to 0.002") in diameter.

You'll need a flow bench and some way to adapt the carburetor so that the carburetor can be flow tested one bore at a time. The rest is just step by step easy. Here we will use a target air/fuel ratio of 12.2:1 because the application is "track race gas" and that will be a safe target. Although the technique can be applied to any flow bench, the following steps are for use on a typical flow bench:

**Step 1:** Remove the float bowls and blow residual fuel (if any) away with shop air or let air dry. The flow bench should not have anything flammable introduced to it.

**Step 2:** Attach metering body with gasket (using spacers) with bowl screws on primary and secondary side of carburetor. Remove power valve if used and use tape to plug off pvcr (power valve channel restrictions).

**Step 3:** It is a good time to verify the secondary opens to WOT (wide open throttle) by bending the secondary link (if used).

**Step 4:** Place the carburetor on the carburetor flow test adapter and install on flow bench. Most carburetor test kits use an additional vertical manometer for measuring signals.

**Step 5:** Place the throttle in the WOT position and secure there for test.

**Step 6:** The vertical manometer is connected to the main jet area (so that the measurement is through the main jet well and includes the full circuit).

**Step 7:** Turn on the flow bench and adjust the flow bench to read 8.2 in.$H_2O$ (this corresponds with 12.2:1 A/F ratio) on the auxiliary vertical manometer. Read the flow: In this example, the primary flowed 77.175 cfm.

**Step 8:** Multiply the flow at the 8.2 in.$H_2O$ signal times 8.2%. (77.175 x 8.2% = 6.32835)

**Step 9:** Refer to the fuel sheet chart on page 148 and look for a number closest to that found in step 8. (In this case, 6.319 refers to a 71 main jet).

Repeating steps 1 through 8 for each of the carburetor venturi is required in order to find the jet that each should run. On the secondary side of the carburetor, the jet size ended up being a 74 main jet (82.83 x 8.2% = 6.792) and the nearest number on the fuel sheet chart is 6.709.

# Flow Testing Carburetors, Fuel Injection & Manifolds

*PVCR Effect*—What about that power valve channel restriction (PVCR) effect on main jet sizing? Because this carburetor uses a power valve in the primary, the main jet size needs to be reduced (at WOT, the power valve is open and adding fuel to the main jet circuit) some amount. The way to find out what size main jet is required when using a power valve is as follows:

**Step A:** Measure the power valve channel restriction and calculate the area. (In our example the PVCR measures 0.039" and the area is $= 0.039^2 \times 0.7854 = 0.0011945$ (PVCR Area)

**Step B:** Divide primary airflow by 100 ($77.175 \div 100 = 0.77175$)

**Step C:** Divide 8.2 by the answer in Step B (in this example, $8.2 \div 0.77175 = 10.625$, this is referred to as a booster-to-venturi ratio)

**Step D:** Divide result in C by 100, multiply times the PVCR area and add to the PVCR area ($(0.0011945 \times 0.10625) + 0.0011945 = 0.0013214$

**Step E:** Divide the result in Step D by 0.7854 so: $0.0013214 \div 0.7854 = 0.0016824$

**Step F:** Take the square root of Step E. ($\sqrt{0.0016824} = 0.041017$). This is the effective size of the PVCR with the draw on the system.

**Step G:** Take the drill size of the main jet (found in determining the main jet without the power valve in Step 8 on the previous page to be a 71 with a drill size of 0.075") and multiply times 1000. In this example, $1000 \times 0.075 = 75$.

**Step H:** Multiply the effective area found in Step F by the answer in Step G (in this case it is $0.041017 \times 75 = 3.076275$. Whew, now we finally have a factor to use!

**Step I:** Subtract the factor found in Step H from the main jet size found in Step 8 on the previous page. In this example, $71 - 3.076275 = 67.923725$ rounding off, this is 68. So at long last, the main jet to run with this tested unit is a #68 in the primary (using a power valve) and a #74 in the secondary.

**Note:** This procedure has worked very well in many, many applications. It is important to use real numbers and not guess at the values, and of course do the arithmetic correctly. Each time you want to change the fuel requirement base, when using a power valve circuit, the entire section must be recalculated. Do not take for granted that the booster-to-venturi ratio is constant. Measure it and plot it so that you can analyze it.

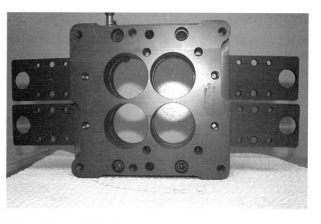

This special carburetor adapter for flow benches is made by Brezezinski and allows each bore to be tested individually or in any combination. Photo courtesy Ranken Technical College.

## Flow Bench Techniques—More Carburetor Testing Methods

For flow bench models using automatic controls, motor controllers and flow computers, the procedures are very specific and must be followed carefully to get reliable data.

Carburetors have sometimes been referred to as "organized leaks," and perhaps one of the many ways to improve the engine is for better organization of the fuel and air circuitry of the carburetor.

Anyone can grind, machine, and modify the carburetor body and/or circuitry. How well the modifications match the engine's fuel requirements is often a challenge and specific measurements and evaluations on a flow bench can provide much needed information.

The first information on flow testing carburetors was supplied to the aftermarket by Holley Carburetors. Carburetor testing on small flow benches and those early instructions set the standard for carburetor evaluations for more than 20 years. A natural extension was for users to test on the larger commercially manufactured benches. Some early large capacity flow benches were initially built for carburetor development. Larger capacity flow benches make testing carburetors much easier, however.

Carburetor testing with a flow bench that is equipped with a flow computer is much easier than older methods using normal manometer pressure measurements if the flow computer is correctly calibrated with the flow bench, but is a bit tricky to do correctly.

## Systemic Approach

The carburetor is part of the fuel and air system controlling the engine and as such must not be viewed as just a component, but must work in conjunction with the overall system. As an example, something so simple as selecting a fuel delivery pressure that is too high will change the characteristics of even the best carburetor.

Fuel delivery, fuel line sizing, selection of the needle and seat, float settings, fuel pressure (while flowing fuel), booster signal variations, flow differences, even mechanical vibration, any and all of these variables are all critical in making a system function correctly. When installed on a vehicle, the vehicle air and fuel system becomes part of the carburetor tuning system.

If you're interested in working more with carburetors, there are a number of excellent books written on the subject. Start with the basics so you understand how the things work.

## Carburetor Flow Ratings

Simple carburetor testing is comprised of such things as measuring the overall airflow capacity of the carburetor body/throttle plate. Collecting the airflow data in cfm (cubic feet per minute) versus some test pressure easily does this type of testing.

What test pressure is a standard for carburetor testing? There is no real industry standard per se, but many modifiers apply numbers from carburetor manufacturers such as Holley and others. Carburetors manufactured by Holley and others are generally rated at one of two specific test pressure numbers. Four-barrel carburetors have been traditionally rated at 1.5 in.Hg and two-barrel carburetors are rated at 3.0 in.Hg. The reason was quite simple. The original Holley company flow benches ran out of capacity at vacuum settings of greater than 1.5 in.Hg (20.4 in.$H_2O$) on the larger four-barrel carburetors, so the number of 1.5 in.Hg was adopted as the standard for their four-barrel carburetor airflow references. Before this method, square inches of venturi area was the method used to rate carburetors. A flow number at a known test pressure is a better way for making comparisons and establishing ratings.

Normal test pressure references on a flow bench are in in.$H_2O$. Note that the two references for carburetors as listed previously are 3.0 in.Hg = 40.8 in.$H_2O$ and 1.5 in.Hg = 20.4 in.$H_2O$. Holley flow numbers are typically listed at "wet" ratings and allow for a given air/fuel ratio (typically 12.5:1) which equates to about an 8% reduction in cfm when considering gasoline as fuel. What this means is that if you measure the airflow of a carburetor at 700 cfm (at 20.4 in.$H_2O$), it would have an equivalent "wet" rating of 644 cfm. This "wet" rating is of course for gasoline. If methanol (methyl alcohol) was the fuel, the reference value would allow for an air/fuel ratio of 5.5:1, which would equate to a value of about 18%. If one used nitromethane as the fuel with an air/fuel ratio of 1.7:1, this would equate to taking up almost 60% of the cfm space. Normally it is easier to relate to gasoline as it is a more common fuel baseline for carburetors. Most needle and seat assemblies might not be compatible with methanol or nitromethane mixtures and are generally too restrictive to fuel flow anyway. But don't let that discourage you from getting creative about solving the fuel flow restriction problems if you want to run copious amounts of non-gasoline fuel through the carburetor.

However, having stated all the above, it is suggested testing the carburetor on the flow bench and allocate for dry testing only as it simplifies (speeds up) the evaluation techniques. Carburetors are all about ratios and signals to provide good fuel curves. Many tests should also be done at part throttle for verification of how well a carburetor can "signal up" as the transition from part throttle to full throttle (WOT) occurs.

A good reference is to test all carburetors (regardless of the number of barrels or throttle bores) somewhere between 15 in.$H_2O$ and 28 in.$H_2O$. One of the many reasons is that some signals within the carburetor are at ratios of 2:1 or more. As you use any test pressure, the booster signal might be twice the test pressure or more. The limitation on some of the commercial flow computer terminals that can be affixed to flow benches is a maximum of 80 in.$H_2O$.

An easy comparison to Holley four-barrel carburetors flow numbers is at 20.4 in.$H_2O$, so I suggest that is an easy reference number to use, although it might be more beneficial to test everything at the same test pressure whether it is cylinder heads, carburetors, or whatever. I have also previously suggested that most things in the airflow path be tested and rated at 28 in.$H_2O$. It will greatly simplify the analysis and the arithmetic for comparisons.

This is far from an official standard, but an easy reference nevertheless. Another approach is to make your own standard of comparison. Whatever number you choose to use, it is important to continue to collect data at the same test pressure reference so it is also easy to compare data in the future. Make lots of notes or you will get lost because you won't be able to remember all the details.

## Comparison Tests

One test that is very valuable to graph is one where the booster signal is compared to various test pressures or different airflow (at different throttle positions). The application of this data is critical in the final analysis of how a carburetor will handle (prepare) the fuel and air mixture that is presented

# Flow Testing Carburetors, Fuel Injection & Manifolds

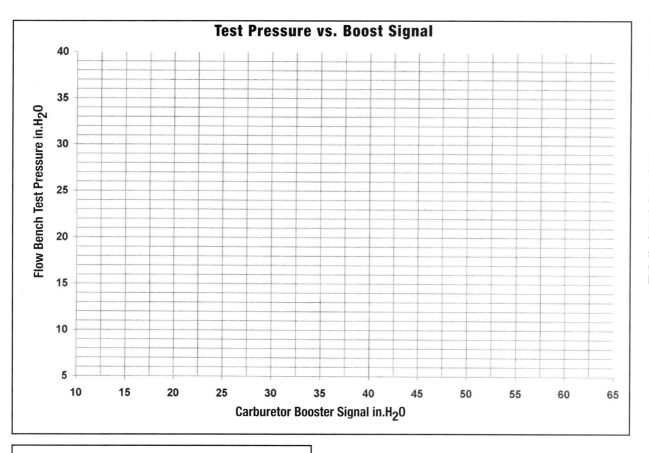

This blank graph can be copied and used for your own carburetor and booster testing. Even if you make your own U-tube manometer for measuring the booster signal, the data will be useful. An ideal circumstance is when all the boosters have the same signal strength with no glitches or hiccups.

## Booster Signal Reference
## Holley Carburetors

| Booster Type (750 DAP) | Test P Flow Bench | Booster Signal in.$H_2O$ |
|---|---|---|
| Straight | 10 | 19 |
| Standard dog leg | 10 | 21 |
| Annular Booster | 10 | 25 |
| Super Booster | 10 | 9.5 |
| Truck Super Booster | 10 | 40 |

| Booster Type (850 DAP) | Test P Flow Bench | Booster Signal in.$H_2O$ |
|---|---|---|
| Straight | 10 | 12.7 |
| Standard dog leg | 10 | 14.5 |
| Annular Booster | 10 | 17 |

| Booster Type (1150 Dominator) | Test P Flow Bench | Booster Signal in.$H_2O$ |
|---|---|---|
| Standard | 10 | 22 |
| Standard 8896 | 10 | 18 |
| Slotted | 10 | 25 |
| Super Booster | 10 | 40.5 |

These booster signal data were shared with me by the late Brad Urban after his Carburetor Shop had done some tests for a magazine article. Brad always wanted to improve his products. Note that the tests were done at only 10 in.$H_2O$ test pressure on a 600 cfm flow bench.

to the engine through the intake manifold.

Another test that is worthwhile to look at graphically is the flow per carburetor venturi/throttle bore. When plotted against different test pressures and booster signals (simulating an engine at wide open throttle undergoing acceleration), the graph can help to identify various and important characteristics of a carburetor.

## Using Flow Computers on Flow Benches

Using flow computer test instrumentation is initially a little confusing, but it can be a very valuable tool for flow testing, and especially carburetors.

The manufacturer's instructions should cover all aspects of how to use the flow computer instrument, but when testing products such as carburetors, some simple procedures need to be applied particularly if the manufacturer did not list how to use the device with carburetor tests.

The flow computer can be easily configured so that the carburetor booster signal is measured by connecting the appropriate signal point with small plastic tubing over or into the main booster channel on the carburetor body. Measuring signals in the carburetor and the numbers generated will be the same with only an arithmetic sign change. Because the carburetor is measured with the flow bench in the intake direction, the correct pressure tap location will provide the correct number including the correct sign. The maximum signal that can displayed on some flow computers is about 80 in.$H_2O$, so be aware of any equipment limitations.

## Booster Balance Test

Some method must be used in order to compare the various booster signals. The easiest method to use is to measure the booster directly and compare to a standard of flow.

Collecting the data and making a comparison graphically is a very informative way to view any individual carburetor and booster package.

Different tests can be done for booster balance tests, such as varying the test pressure (while the throttle is at WOT), or with varying the throttle position and maintaining the same test pressure. The latter method requires a very accurate reference for throttle position.

## Jet Size Based on Calculations and Flow Bench Testing

Some simple calculations will provide a very close reference main jet size for initial trials on carburetors. It should be obvious that the correct jet and resultant fuel curve should be proportional to airflow through the carburetor venturi(s).

The selection of the main jet and the selection of air bleed sizes are very closely related, but also are dependent upon the proper fuel delivery supply for the system to perform up to expectations.

$V_2 = \sqrt{[((2g(P_1 - P_2) \div w_1)) \div (1 - (A_2 \div A_1)^2)]}$
where $V_2$ = velocity of air ft/sec, g = acceleration of gravity ft/sec/sec, $P_1$ = absolute pressure prior to venturi in lb/sq ft, $P_2$ = absolute pressure at venturi in lb/sq ft, $w_1$ = weight density of air in lb/cubic foot, $A_1$ = area of entry cross-section in square feet, $A_2$ = area of venturi section in square feet.

Essentially the main jet should be at some ratio to the main and booster venturi diameters. And an easy way to toss a crude estimate at a carburetor is by applying the following equation:

$M_{jd} = V_{dc} \div 19$
where $M_{jd}$ = main jet diameter in inches, $V_{dc}$ = carburetor venturi diameter in inches. Then you have to see what the equivalent jet number is after calculating the diameter of the main jet.

The Holley jets are problematic in that they have many jets that use the same diameters and flow is changed by altering approach angles and departures. This method is very crude and does not take into consideration any variables for weather or fuel changes or those issues. As stated it is just a pretty crude estimate, but it is quick.

If you had a carburetor with a main venturi diameter of 1.375" then the quick calculation would show that you should select a main jet diameter of 0.072". That is very close to a Holley number 70 main jet (actually 0.073" diameter).

*Simple Math Trackside Tuning*—The first barometric pressure readings were done in 1643 by Italian physicist and mathematician Evangelista Torricelli (1608–1647) when he invented the barometer. He had no idea at the time of how important his work would be to any racer or dyno tuner. Somewhat humorous, how that worked out, because his work was targeted to study water pumps of that time.

During dyno or racetrack testing, suppose that conditions were barometric pressure reading of 29.75 in.Hg and dry air temperature was 80°F. And wet bulb temperature was 68°F. The tests went well and the engine made good power and the tuned A/F ratio was 12.5:1 using a #78 main jet in the Holley carburetor.

You load the vehicle up and head to the racetrack that is over 150 miles away and the following weather conditions are prevalent. Trackside barometer is a solid 28.90 in.Hg and the dry bulb temperature is 85°F. You check the water in the air with your psychrometer gizmo and the reading is 72°F wet bulb. What jet should you start with? For the sake of clarity, we will assume that this example vehicle does not have an air scoop in place that would add pressure to the carburetor inlet. This is a fairly conservative approach because the under hood temperature will be higher than the ambient dry bulb temperature and thus will allow the mixture presented to the carburetor to be slightly richer than the calculation will show. This calculation standard is for 29.92 in.Hg, 60°F, dry air so at these conditions, the number is 100%.

Jet = 100 x $\sqrt{(29.92 \div P_b - V_p)}$ x $\sqrt{T_{air} \div 520}$
where $P_b$ = local barometric pressure in in.Hg, $V_p$ = local vapor pressure in in.Hg, $T_{air}$ = inlet air temperature°F + 460.

If the calculated number is greater than 100, the reference is that the target mixture influenced by the main jet would be leaner. If the number were less than 100, it would indicate the number is richer.

# Flow Testing Carburetors, Fuel Injection & Manifolds

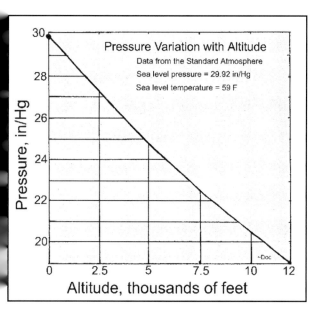

This graph is useful for track tuning or keeping in mind that the atmosphere is the basis for supplying atmospheric oxygen to the engine. Get a grip on what atmospheric effects have on the engine and deal with them accordingly.

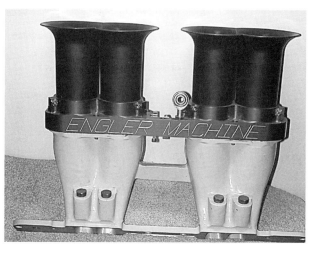

These injectors are about as nice as they get. The Engler system has gained lots of popularity over the last several years. Shown here prior to bolting on to a cylinder head for airflow testing.

Baseline data calculation:
$$\text{Jet}_{base} = 100 \times [\sqrt{(29.92 \div 29.75 - 0.57)} \times \sqrt{540 \div 520}]$$
$$\text{Jet}_{base} = 103.19$$

Trackside data calculation:
$$\text{Jet}_{track} = 100 \times [\sqrt{(29.92 \div 28.9 - 0.62)} \times \sqrt{545 \div 520}]$$
$$\text{Jet}_{track} = 105.3$$

The difference ($\Delta$) here is 105.3 − 103.19 = 2.11 so the mixture should be leaner than the baseline jet by 2.11%. Holley jet numbers are supposed to change the fuel flow by about 2% per jet number change (according to Holley information). The difference in trackside conditions vs. baseline conditions indicates that you should change the main jet by one number smaller than at your dyno or track testing sessions. So the new jets would be #77s. That is if they were all the same. Times like these make you want to convert to EFI because it is easier to use the laptop.

However, there is some general question about how close the tolerance of the jets is to the 2% per Holley number listing. Some very accomplished racers have chosen to use the MaxJets from CompCams because they are available in 0.001" diameter increments. If you were using the MaxJets, then the jet change to lean the mixture would be a reduction of the area of the jet by about 2%. That is the reduction of the area, not the diameter of the orifice. Simply remember that the area can be calculated by using the following equation:

$$A = D^2 \times 0.7854$$
where A = area in square inches,
D = diameter in inches.

## Testing Throttle Bodies

Testing throttle bodies is largely a problem of flow bench capacity at WOT. Some of the popular throttle bodies for electronic fuel injection (EFI) systems have huge airflow capacity. The testing can be carried out at partial throttle positions in order to gather data for good input to a mapping program that relates to the TPS (throttle position sensor). Some TPS units go to the same as WOT at anything over about 60% on the TPS output.

Mechanical fuel injection systems can also be tested at partial throttle and at WOT conditions on the flow bench. The mechanical fuel injection system manifolds and throttles are best tested while connected to a cylinder head because that is the way the system will run. It doesn't take too long for a tester to realize that just having large throttle blades does not make the system flow more. Again, it is shape and not size that becomes a deciding issue for these type units. There is also some advantage to having the throttle valves (blades) up away from the cylinder head intake port so that the air has some space to recover from the passage past the throttles.

## Testing Mechanical Fuel Injection

The things that should be considered in mechanical fuel injection systems are that the fuel system and the air systems generally share the same eventual flow path into the engine's cylinders. The testing methods for the airflow side should be done very similar to the description for throttle bodies previously listed.

# Engine Airflow

These old-style (purchased new in 1965) small Hilborn injectors were rebuilt and will be used on an old engine for Bonneville. The first efforts will be on gasoline and switching to methanol then Nitromethane. Yes they were tested on the flow bench with the cylinder heads and with radius inlet tubes. They flow a bit more than the cylinder heads do.

The airflow path of most stock castings or machined throttle housings can be improved by carefully smoothing transitions from machined portions to cast portions. It is also pretty common for the port location to not be in the best place either.

If there is port misalignment, take a look at improving it and test it on the flow bench before and after. Winners are built by hard work, not by magic or wishing for perfection.

The fuel system side is a little more difficult, but is outlined here for the reader's convenience. It is just another way to look at the problem so that you can have an appreciation for how to solve some of them.

## Analyzing Fuel System Requirements for Mechanical Fuel Injection

The fuel requirement of a high performance engine can be predicted based upon a simple analysis of required power Vs fuel type. The power inputs into the analysis can be either from actual testing or estimated from a target performance number.

The following words are not intended to replace the documentation and instructions that manufacturer's supply with each mechanical injection unit. Details such as placement of springs, check valves, and procedures for adjusting the injector unit are omitted from this description and it is strongly suggested that the instructions from the original manufacturer be used as the best guideline for proper operation.

*Basic Performance Calculations and Estimates*—The power requirement for quarter-mile performance can be calculated by using the formula:

$$hp = (0.00426 \times mph)^3 \times Veh\ Wt$$

where hp = horsepower, mph = vehicle terminal speed in quarter-mile, Veh Wt = vehicle weight in lb.

A pretty good estimate for power at the tire patch would be to use the following formula:

$$hp_{est} = (0.0040 \times V_v)^3 \times W_v$$

where $hp_{est}$ = estimated horsepower (tire patch), $V_v$ = quarter-mile trap speed of vehicle in mph, $W_v$ = vehicle weight in lb.

*Drivetrain Factors*—Because there are various losses in a drivetrain, in order to estimate flywheel power, some number has to be applied in order to estimate a fuel requirement.

A good estimate for most drag racing vehicles is greatly dependent upon transmission type. Automatic transmissions are fairly efficient these days, but the fluid coupling that is also called a torque converter is generally very inefficient. The slippage certainly can be a problem particularly when "high stall" racing type torque converters are used. Suggested numbers to provide initial evaluation of total drivetrain losses are:

Automatic transmission with racing or high stall converter can have losses of 100 hp to 200 hp (normal losses average 90 hp to 100 hp).

Manual transmission drivetrains with properly adjusted clutch have typical losses of 60 hp to 100 hp.

I strongly suggest you use an average loss number for each application and evaluation as an initial power number for planning purposes. Remember that these numbers are estimates (albeit very close estimates).

## Liquid Fuel Types

The type of fuel used greatly affects the fuel system planning and performance estimates and calculations.

The specific gravity (SG) of water ($H_2O$) is 1.00, weight in lb/gal at 68°F = 8.34 lb/gal. The SG of methanol ($CH_3OH$) is 0.794, wt lb/gal at 68°F = 6.62 lb/gal. The SG of ethanol ($C_2H_5OH$) is 0.785, wt lb/gal at 68°F = 6.55 lb/gal. The SG of gasoline ($C_7H_{17}$ is typical) varies greatly, so in order to find the lb/gal, multiply the SG of the sample times the weight of $H_2O$ (8.34 lb/gal). The SG of pure nitromethane ($CH_3NO_2$) is 1.139, weight in lb/gal at 68°F = 9.5 lb/gal.

Nitromethane is used in varied percentages in methanol, so the SG of the percentage mix should be multiplied times the SG of $H_2O$ (8.34 lb/gal). For initial calculations and planning, you can use the chart of nitromethane percentages in methanol, at right.

## Brake Specific Fuel Consumption for Different Fuels

The BSFC for various fuel types is listed below for use in calculations and estimations for fuel system planning. In general, BSFC is expressed in lb/hp-hr.

The BSFC for gasoline ($C_7H_{17}$) is approximately 0.5 lb/hp-hr.

The BSFC for methanol ($CH_3OH$) is 1.00 to 1.5 lb/hp-hr. Engine and piston cooling is greatly affected by using very rich mixtures. Normally aspirated engines run closer to 1.04 lb/hp-hr in typical field applications. Highly supercharged engines run closer to 1.25 to 1.5 lb/hp-hr or more. Many successful highly supercharged engines running 40+ psi boost run methanol BSFC numbers at 1.8 to 2.2 lb/hp-hr with much of the fuel being used as an additional coolant as it passes through the engine. Those highly boosted engines are sending loads of fuel through the engines.

BSFC for nitromethane depends upon the percentage and engine configuration.

A good number to use for planning is somewhere between 0.7 to 0.8 lb/hp-hr (for nitromethane at 40% to 90% in methanol). Lower percentages on naturally aspirated engines can use a BSFC number of 1.0 lb/hp-hr as a starting point for evaluation.

*Fuel Injection System Type*—It is imperative that you have a good working knowledge of the fuel injection system used for accurate analysis and planning. Details such as various pressures Vs flows, bypass operation, nozzle bias, and pump performance should be known. More complex details might be the pressure and flows a second set of nozzles might be activated.

If specific flow Vs pressure numbers are not known, then consider using very conservative or higher engine BSFC numbers for the fuel type. When the specific flow Vs pressure curve is not known (by measurement), use the manufacturer's information on pump output as an initial point for planning.

It is common that some systems restrict the fuel to the barrel valve and the nozzles while some restrict the flow to the tank on the return fuel side. Each type scheme has its own applications, so make no assumptions and learn how the mechanism works before you tackle one of those projects.

### Nitromethane in Methanol ($CH_3NO_2$ in $CH_3OH$)

All chart listings are for 68°F and specific gravity (SG) is for percentages as listed. The additional column of data on percentage of nitro is a reference for volume numbers.

| % Nitro (by weight) | S.G. (@ 68°F) | % Nitro (by volume) | ~A/F ratio (LBT mixture) |
|---|---|---|---|
| 0% | 0.792 | 0% | 5.00:1 |
| 5% | 0.809 | 3.53% | |
| 10% | 0.827 | 7.17% | 4.96:1 |
| 15% | 0.844 | 10.90% | |
| 20% | 0.861 | 14.80% | 4.80:1 |
| 25% | 0.879 | 18.80% | |
| 30% | 0.896 | 22.90% | 4.53:1 |
| 35% | 0.913 | 27.20% | |
| 40% | 0.931 | 31.60% | 3.39:1 |
| 45% | 0.948 | 36.20% | |
| 50% | 0.966 | 41.00% | 2.71:1 |
| 55% | 0.983 | 45.90% | |
| 60% | 1.000 | 51.00% | 2.26:1 |
| 65% | 1.018 | 56.30% | |
| 70% | 1.036 | 61.80% | 1.94:1 |
| 75% | 1.052 | 67.50% | |
| 80% | 1.072 | 73.50% | 1.69:1 |
| 85% | 1.087 | 79.70% | |
| 90% | 1.104 | 86.20% | 1.50:1 |
| 95% | 1.122 | 92.90% | |
| 100% | 1.139 | 100.00% | 1.35:1 |

**The increase in power with the addition of nitromethane in methanol is fairly linear until about 80% or so. Nitromethane is very prone to detonate and is very sensitive to compression ratio (both mechanical and dynamic) as the percentage passes approximately 25%. Larger percentages typically take more ignition timing because the mixture is harder to ignite. It is not slow burning and when the mixture is ignited and combustion is supported, the burn rate is faster than methanol-only fuel because of the additional oxygen content. Nitro mixtures continue to burn when the exhaust valve opens as the exhaust pipes of fuel-burning race cars clearly demonstrate. The "feathers" of nitro flames will capture your interest and sing a song of attraction much stronger than that of gasoline.**

**The thing to bear in mind when using nitromethane is to run the tune-up very rich. The A/F ratios listed above are for reference only and it is recommended that the actual tune-up used on the racetrack should be much richer. The listing of LBT (lean best torque) is often replaced with RBT (rich best torque) and is preferred to make maximum reliable horsepower.**

**The table above is provided for a comparison only and for general references for naturally aspirated engines as supercharged engines have higher dynamic compression ratios and typically will require much more fuel to perform well. In reality, one cannot get 100% nitromethane to combust properly and it is much more common to dilute the 100% by use of small percentages of methanol or benzene (preferably spectrophotometric benzene).**

# Engine Airflow

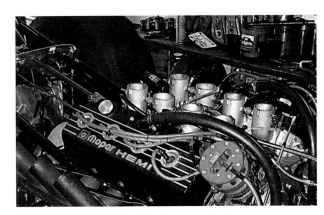

Hilborn injectors on a Hemi are pretty natural around drag racing. They are easy to tune once you understand how the mechanism works. The radius entry ram tubes were removed here to check the throttle settings at idle (idle air) but should be used because they improve airflow into the injector body.

## A Real-World Example

For a practical example of fuel requirements, let's consider a drag racing vehicle weighing 3,000 lb. (with driver). The vehicle is equipped with a Powerglide transmission and a 4000 rpm stall converter. The target quarter-mile trap speed is 170 mph. The engine is supercharged and the fuel is methanol. The injector is an Enderle unit using a standard bypass and the pump is an Enderle 110. What is a good starting point for the injection system? What is an approximate predicted ET based on the power?

$$hp = (0.00426 \times 170\ mph)^3 \times 3{,}000\ lb. = 1139\ hp$$

This is a reference for power at the flywheel:

$$hp_{est}\ @\ flywheel = (1139\ hp + 125\ hp) = 1264\ hp$$

Assuming a good run with adequate traction at the tire patch, 170 mph trap speed should produce an ET of about 8.02 seconds.

Using a BSFC number of 1.2 lb/hp-hr yields 1517 lb/hr fuel flow at peak power.

$$1517\ lb/hr \times 1\ hr/60\ min \times 1\ gal/6.62\ lb.$$
$$= 3.82\ gal/min\ required\ at\ engine\ demand$$

Using a factory-supplied pump curve, the Enderle 110 pump capacity is referenced at 13 gpm @ 4000 pump rpm (8000 engine rpm).

***Bypass Size***—So, the engine is to use 3.82 gpm and 9.18 gpm should be returned to the fuel tank. What bypass size should be selected? From a chart from the manufacturer, or by calculation, the bypass jet diameter should be 0.190". This jet size will provide return fuel flow of 9.10 gpm at 100 psi (pump pressure at 8000 engine rpm). What about the remaining fuel supplied in excess to the engine?

1. A high-speed bypass can be used to return additional fuel at a selected rpm and pressure when the pressure curve of the fuel system is known.

2. The engine can be run in an over-rich condition (methanol is very forgiving) without many negative consequences. Selection of components to provide a rich condition is much wiser than choosing the lean side, which is not forgiving.

So what size high-speed bypass and at what pressure? Selecting about 65 psi will yield an engine speed of about 5000 to 5500 rpm. The high-speed bypass jet that will accomplish the target is only 0.030" in diameter.

It should be obvious that a different combination could also be provided so that the initial bypass could be sized to be a smaller diameter, and then select a larger high-speed bypass jet with the same overall flow control. It will just change the curve and where it might be changed.

***Nozzle Size***—What size nozzles should be installed to provide fuel flow to the engine the targeted 3.82 gpm as outlined above? Using either a manufacturer's guide or by calculation, 8 nozzles of 0.044" diameter can be used if all the fuel is directed through the topside of the supercharger at 100 psi. The nozzles do not operate at 100 psi throughout the range, so an oversize of the 0.044" size would be in order. A selection of nozzles that are 0.052" or 0.055" will provide more than enough fuel to the engine at only 50 psi. This is why the pump supply curve is so important.

The best selection process is of course to liquid flow test (using a liquid flow bench or test system) the complete system including all the lines to be used on the race vehicle.

***Data Needed for Analysis***—Pressure and flow data should be very accurate in order to ensure adequate results. Necessary data to obtain for any system includes: Pump output curve and raw data for flow vs. pressure, pressure at the pump outlet, pressure at the nozzles (includes any auxiliary applications), pump rpm (so the pump drive ratio is selected correctly). It is always worthwhile to accurately measure the check valve blow-off pressures.

Barrel valve adjustments should be based on a standard leakage percentage target so that changes can be easily compared with a numerical reference.

Also be aware that if the engine is supercharged or turbocharged and using port nozzles, the boost pressure is working against the port nozzle fuel flow.

# Chapter 11
# Selecting Camshafts with Flow Numbers

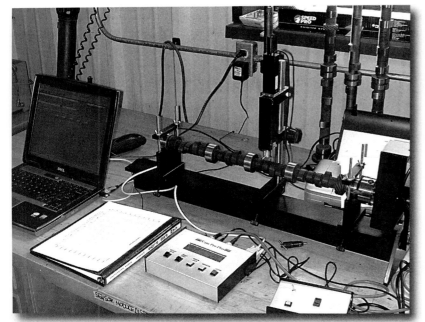

This is a camshaft measurement tool that allows both data and a graph to be stored and displayed on a computer. This unit is the Cam Pro Plus from Audie Technology and can help advanced builders understand more about cams. It can also be used with the cam installed in the engine so it makes the degreeing process a breeze.

*I learned very early the difference between knowing the name of something and knowing something.*
—Richard Feynman

In airflow and engine seminars one of the most asked questions is one that is very basic in engine building. How should a camshaft be chosen? The choice of a camshaft is something that can be done based on airflow data and sharing that information with the camshaft company.

Most knowledge based cam companies staff their telephone tech lines with fairly bright people and want their customers to do well with the suggested camshaft selections. The people that answer the general tech line calls are given a very strong database of successful selections for given engines, but might lack the higher level expertise that a professional engine builder would want to depend upon. When faced with a condition like that, there is probably a tech group within the cam company that is sometimes called the "engine builder department" and is typically staffed with people that have a higher level of engine awareness and can speak more in-depth to the details required.

One thing is for absolute certain and that is you don't need to go immediately to the bottom of the page of the cam catalog and proclaim, "Wow that must be the one!" That type of selection process is a guarantee that you might end up with the most rumbling idle and perhaps more valvetrain maintenance than you had ever bargained for.

## Working the Numbers and Selecting a Camshaft with Flow Bench Airflow Data

How should a camshaft be chosen? The camshaft selection should be based on airflow and thus should be chosen last when choosing complimentary engine components. Notice that the statement said last. Even if you have kept the good old favorite cam with the grind identification ground off so nobody would know how "trick" your bumpstick might be, be patient and try a different approach. Or maybe you already have made a lucky choice without having any data to base it upon.

First, you need to collect the airflow data for the complete intake system on an accurate flow bench. The importance of flow bench accuracy here is that if your data is not accurate, then the camshaft also will lag behind the target of truth that will produce the best results. A few calculations can be done to target some information to narrow the selection for cam timing, particularly at the point of intake valve closure. Choosing the best closing point will be an advantage in maximizing the inertia supercharging for the engine. Of the four events of the valve timing, the intake closing number is the most important.

If this is not clear to you reread the section on valve events and about how air gets through the engine. That was back in Chapter 2.

The amount of flow per square inch of the intake valve (cfm per square inch) vs. crank position is an easy way to help plot the information for use. The use of this data will give the user the capability to plot the data and to calculate the Cv, which is a rating for the complete intake system flow. Note that the cfm/square inch varies with test pressure, as does the potential flow line. Common potential flow line values at 28 in.H2O test pressure is 146 cfm/square inch. You can do your own graph for another test pressure if you

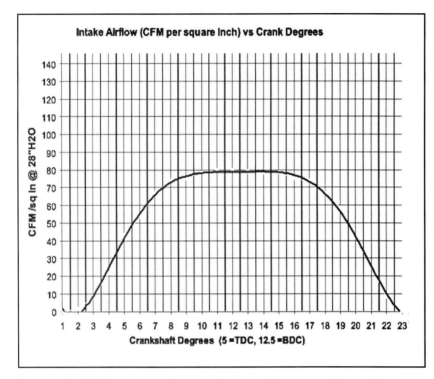

The graph will help you to understand how to plot your own data, and by simply "counting the squares" you can come up with an intake system flow rating. Be patient and it will all make sense in the application of some numbers and a few graphs.

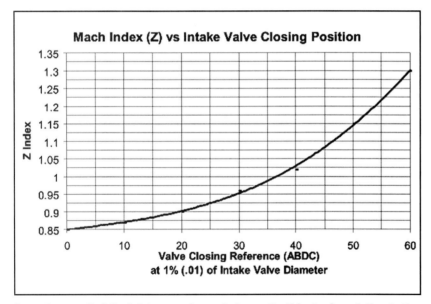

Sometimes called the inlet supercharge index or the Z factor for relating to the inlet Mach index, this graph shows how to rate the inlet package. It takes a little reading and looking, but the efforts will be worth it. This was first used by Taylor in his textbook. The Japanese referenced it as a Mean Inlet Mach Number.

read the earlier text in detail.

$C_v = A_{fc} \div A_{tp}$
where $A_{fc}$ = Area under flow curve, $A_{tp}$ = Total area under the potential flow line, $C_v$ = Intake system flow rating

The $C_v$ on better engine designs will probably be somewhere between 0.40 and sometimes 0.45. The higher the $C_v$ value, the better the engine design. This is not to be confused with the coefficient of velocity. Be sure that you actually read these descriptions before attempting to apply them in any fashion.

Camshaft designers will typically want to consider rpm, airflow, and displacement in order to generate a potential curve of power Vs rpm. This is when they would want to know your data on airflow from the flow bench.

The area of the two curves shown can be found by simply counting the squares under each. When you plot your own data, it is best done on graph paper that has 10 squares per inch which provides good resolution and easy numbers to deal with.

Using the $C_v$ from the previous calculation, the average inlet area can be calculated:

$A_{ia} = C_v \times A_v$
where $A_{ia}$ = average inlet area, $C_v$ = intake system flow rating, $A_v$ = valve area, square inches

The calculations above can then be used to find the intertia supercharge index, $Z_i$:

$Z_i = (rpm \div 126{,}000) \times \sqrt{[(D_{cyl} \times L_i) \div (A_{ia})]}$
where $Z_i$ = inertia supercharge index, $D_{cyl}$ = displacement of 1 cylinder, cubic inches, $L_i$ = length of the inlet, inches, $A_{ia}$ = average inlet area, square inches, rpm = revolutions per minute

Note that the Length of the Inlet, $L_i$, is the length of the inlet tract from the edge of the intake valve to the opening into expansion at the first plenum. The complete length includes the cylinder head port and the runner in the manifold.

The $Z_i$ value for most engines will fall between 0.9 and 1.2. When $Z_i$ is calculated one can use the graph in for planning to close the inlet valve to a point of 0.01 x the valve diameter. This is also referred to as 0.01 L/D or 1% of the diameter of the valve in lift.

Note the closing reference on the curve above would be 0.01 x valve diameter (only 0.020" if using a 2.00" diameter valve). This examination will provide the inlet closing number and the

# Selecting Camshafts with Flow Numbers

This close up shows a camshaft lobe that is definitely not symmetrical (opening and closing sides the same from centerline). The lobes are very noticeable as being asymmetrical (not the same on opening and closing sides from centerline). This motorcycle engine uses a "wiper" follower/rocker arm. An asymmetrical lobe on a pushrod V8 would be more difficult to see unless you were graphing the lift vs rotation.

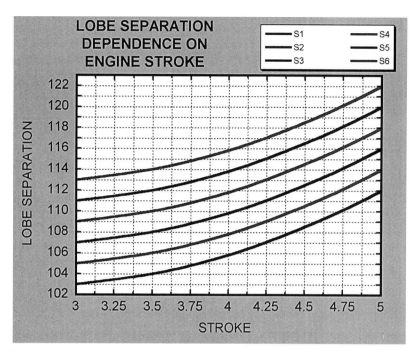

One of many things that can have an effect on lobe centers and lobe separation angles is the stroke of the crankshaft. This is not meant to replace your contact with your cam company, just to help you understand it is not just one thing. Courtesy Comp Cams.

resultant selection of the camshaft and valve timing will follow consultation with a cam company and evaluation of the application in detail.

It is worthwhile to note here that it is probably quite beneficial for the intake valve to be open to a point of at least 0.02 L/D or 2% of the diameter of the valve in lift by at least 15 to 20 degrees BTDC. Valve opening points of up to 40 degrees BTDC are generally much more preferable. These are general guidelines so that the engine can take advantage of positive effects from intake inertia supercharging which is sometimes called intake tuning. These parameters are only of some gain if the Z factor mentioned earlier is at least 0.3 or greater.

Some basic rules of thumb in the choice of cams for a 23-degree cylinder head Chevy will have the lobe separation angle (LSA) generally being between 107 degrees and 111 degrees. Meanwhile the duration (referenced to a lifter rise of 0.050") will generally be between 255 degrees and 275 degrees. Short manifold runners or poor flowing cylinder heads will normally require changes in the LSA and duration (tighter and longer, respectively). Other engines will probably have different characteristics.

Before you pick up the telephone and call a cam company and say that you want a cam based on airflow, carefully think the numbers through. Verify that your numbers are correct so you will understand the manufacturer's recommendation.

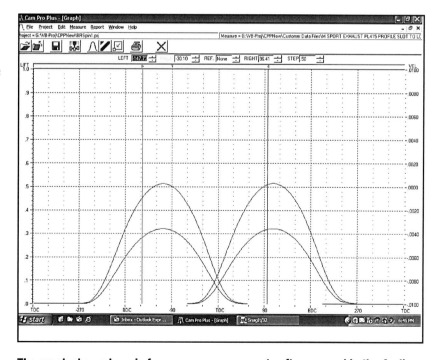

The graph shown here is from cam measurement software used in the Audie Technology CamPro. The lift vs crank position is shown for the cam lobes and the lift at the valve. Photo courtesy Audie Technology.

# Chapter 12
# Graphical Analysis

*In questions of science, the authority of a thousand is not worth the humble reasoning of a single individual.*
—Galileo Galilei

**Countless hours spent testing doesn't do much good unless you know how to analyze the data. As I've said countless times in throughout this book, you have to know where the numbers come from in order for them to have any meaning. Courtesy Mike Mavrigian.**

The use of simple graphs can add much to your capability to analyze and evaluate data. A simple graph is typically done in a two-axis format. The vertical axis is Y and the horizontal axis is X. It is best to scale the graph so that the vertical axis (Y) is flow in cfm and the horizontal axis (X) is valve lift. As with any data, make careful notes so that you can remember what your test was about.

In this chapter are several graphs of different airflow tests. These are by no means all the data available, but are shown for ease of comparison and to get you used to looking at data in this easy to understand format. Obviously not all makes and designs of cylinder head airflow are presented, but these will get you to thinking more about airflow. It is always better to accomplish your own testing or witness a test than to simply believe whatever some dude says was the hot tip.

The data presented is typically listed in columnar format (raw data) and converted to the graphical. You will notice the graphical data is by far easier to take a quick look at instead of trying to think like an accountant looking at columns of numbers.

All the graphs on flow are presented with data taken at 28 in.$H_2O$ test pressure unless otherwise described. Because the graphs are all in black and white format, in general the intake flow will be a solid line and the exhaust flow will be a dashed line unless otherwise signified.

# Graphical Analysis

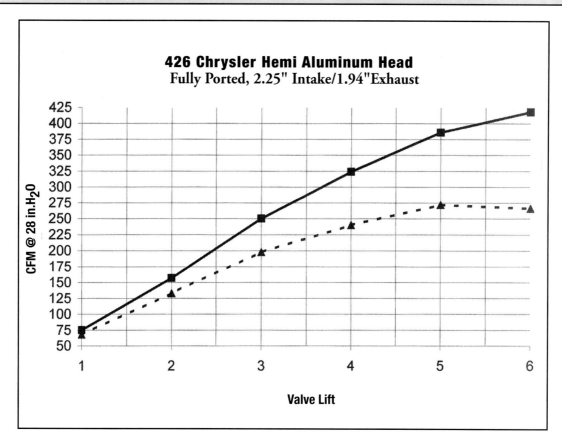

The famous 426 Chrysler Hemi aluminum head when properly reworked flows a lot and as you can see on the graph, it keeps flowing as its valves are lifted higher.

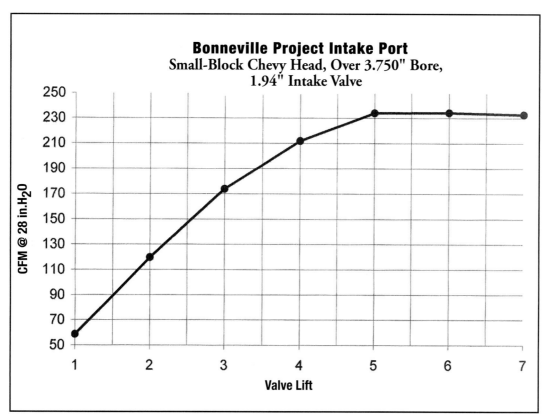

This is an intake port only tested over a 3.750" bore. It is a small V8 that will be used at Bonneville someday. The flow is pretty good for over a small bore with a 1.94" diameter intake valve. Not too shabby for an old reworked iron head.

This one was inspired by meeting a Cadillac guy and encouraged the thoughts of putting a 500 cid Cad into a Chevy Suburban for a tow rig. Should work out to get good mileage.

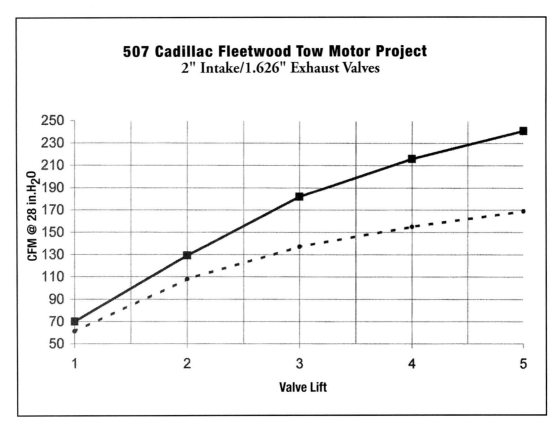

The dashed line is a cast iron 034 casting number small-block Chevy head with some bowl work and a 2.02" intake valve. The solid line is the aluminum small-block Chevy as-cast number 298 with 2" intake valve.

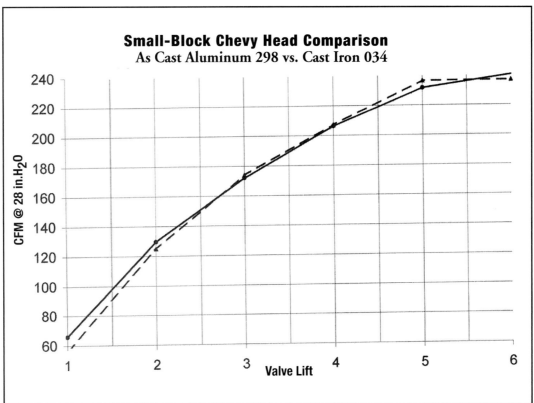

# GRAPHICAL ANALYSIS

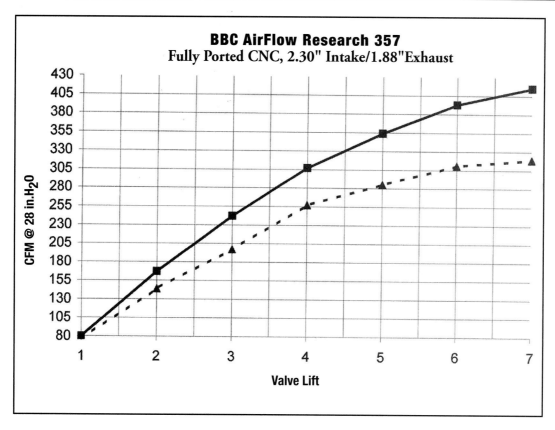

These BBC heads came from AFR right off of the CNC machine and the flow data very closely matched the AFR factory numbers. These heads made well over 1000 CBhp on methanol in early dyno testing on a 540 cid BBC. Good stuff!

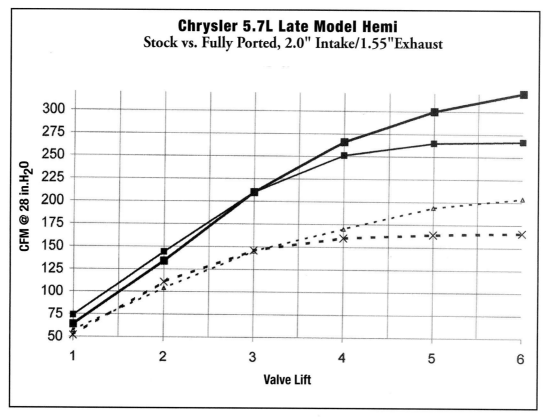

This graph shows a comparison of stock head flow vs. fully ported head flow for a late-model Chrysler Hemi 5.7L engine. Some real potential there.

# Engine Airflow

If your taste is in Mopar engines, this flow test should get your attention. The graph is for a NHRA "legal" Stock Eliminator engine. This is for the 440 cid engine and much effort was put into the valve job and the valve preparation. This is the kind of stuff that drives tech guys nuts.

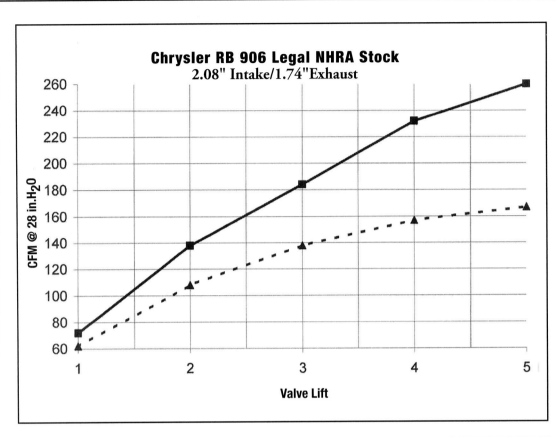

These 440 Chrysler heads were ported and prepped and the work obviously pays off. Getting creative on valve seat approach angles and on the valve shapes is beneficial.

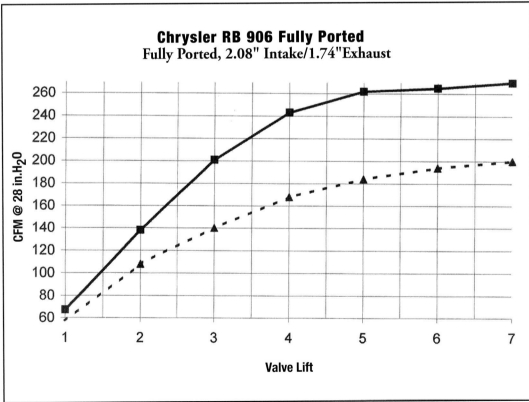

# GRAPHICAL ANALYSIS

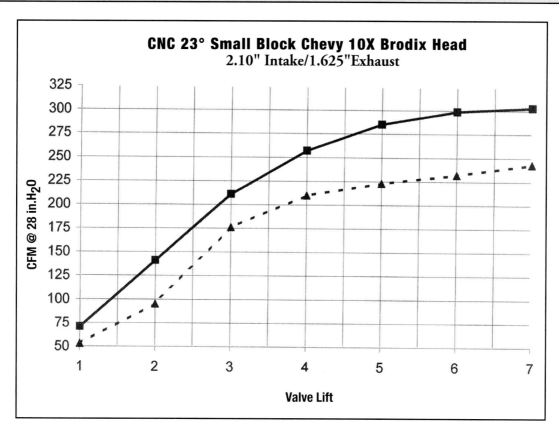

This set of castings was tweaked by hand finishing after running through the CNC machine process. The Brodix heads of this type were very popular with the sprint car guys that used 400+ cid engines.

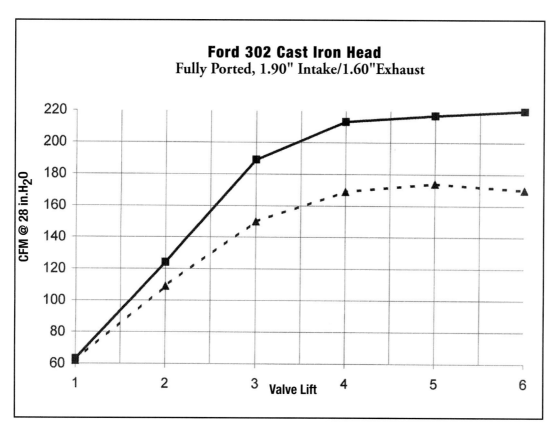

The Ford guys have had a great time with the 302 cid platform but the things have always needed better cylinder heads to perform well. Notice how these fully ported cast iron pieces work. For a quick comparison note the Chevy 034 heads.

Using ported 351 Ford cast iron heads on a 302 cid or even on a 289 cid can be done and the flow graph shows that might be a worthwhile consideration.

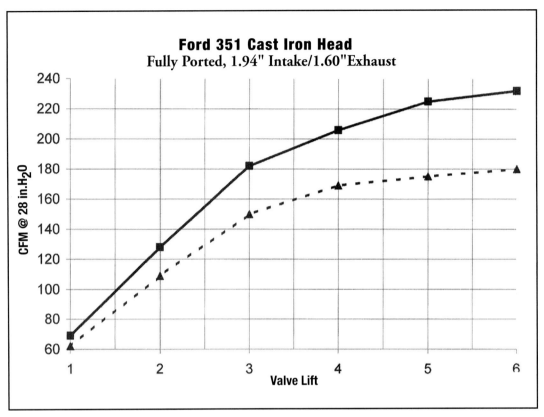

The old Ford flathead was the basic V8 that started the hot rod revolution and ruled from 1932 to 1953 and they are still popular today. The flow graph shows why overhead valve V8s suddenly became an easy choice to make.

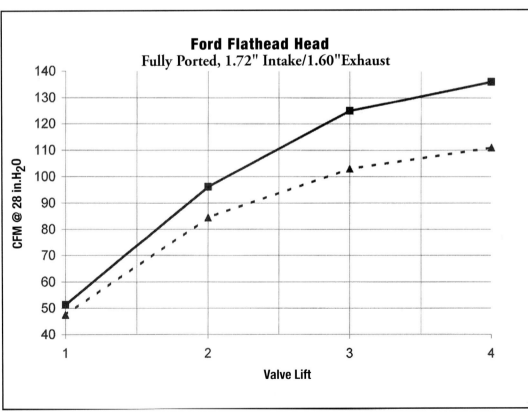

# GRAPHICAL ANALYSIS

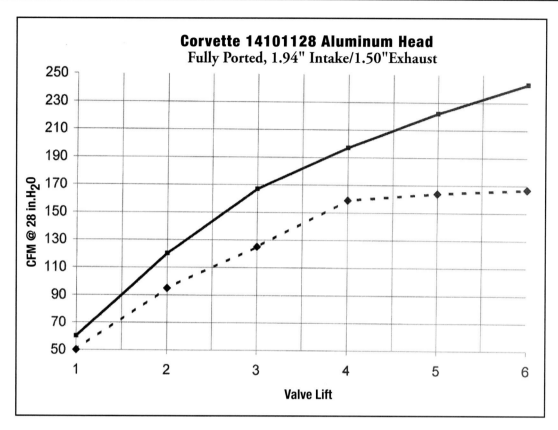

This flow graph shows how impressive a 1.94" intake valve can work when the Corvette head casting is properly reshaped. They were reworked by JD Engineering and might be a good alternative to the iron 492 heads for a Bonneville project.

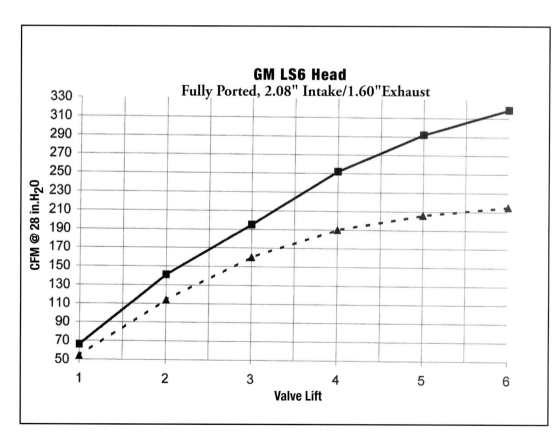

The GM LS series engines have produced some very impressive cylinder heads for gearheads that want to work with them. This flow graph of a fully ported LS head shows why.

**This flow graph shows the effects of less than an hour of work on the intake port bowls. The 455 Olds head has a funny crook in the intake port and as you hopefully have learned by reading this book, air hates to change directions. More time is required for better results.**

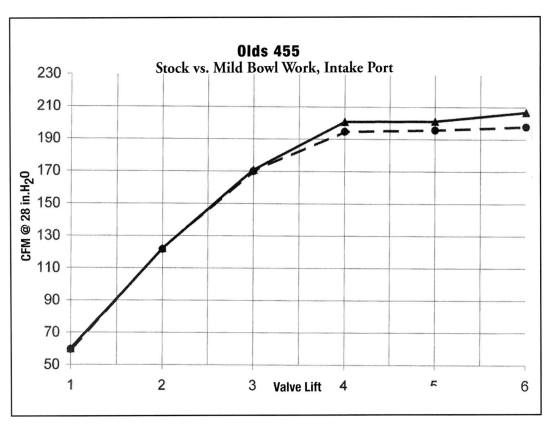

**Harley-Davidson parts have been around forever and many of the components get modified seeking better performance. This flow graph shows the comparison of a stock vs. modified EVO head.**

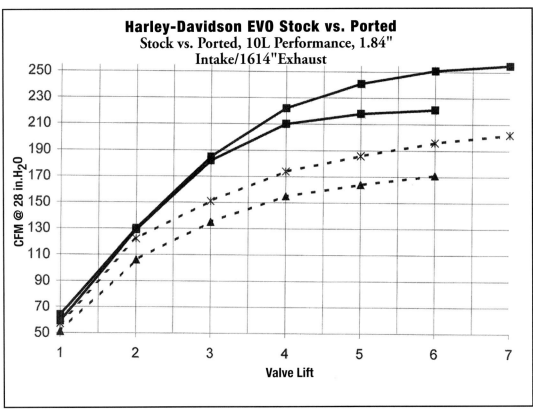

# Graphical Analysis

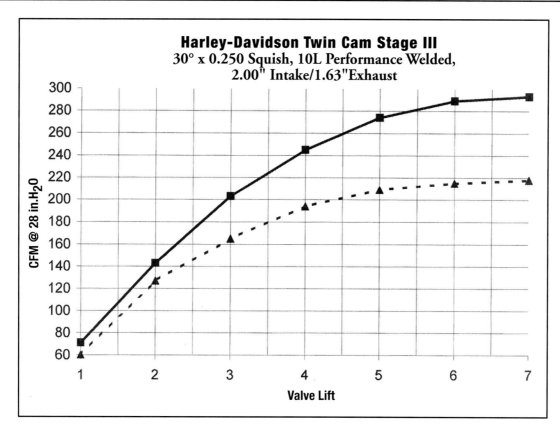

This flow graph shows how effective a welded and reshaped twin-cam H-D cylinder head can be. These things can be made to flow some air but it takes lots of skillful work like this one done by 10 Litre Performance.

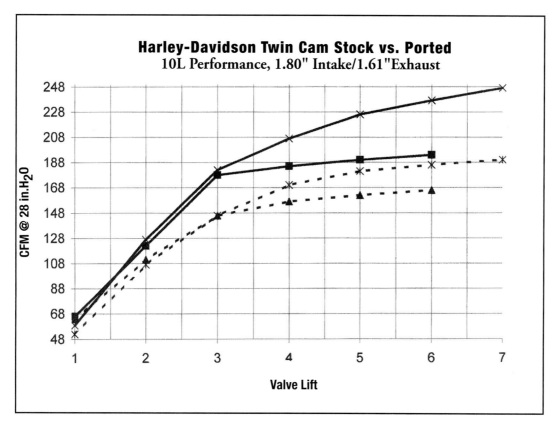

The flow graph shows the difference in stock vs. ported characteristics for a twin cam H-D. These things can obviously perk right up if you know how.

# Engine Airflow

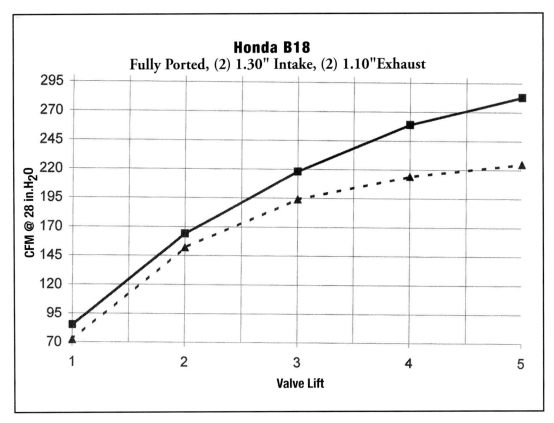

Honda B18 cylinder head fully reworked shows how impressive the 4 valve packages can be. No wonder they respond so well. Even though this graph shows valve lift to 0.500", many of the 4 valve engines only lift to a little over 0.400" (depends on the camshaft setup).

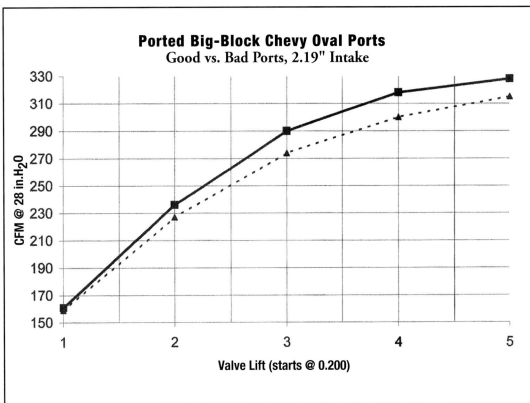

This flow graph shows the difference in flow of the good vs. bad ports in a big-block Chevy oval port head ported by VortecPro. This was a 049 casting. Big-block Chevy heads typically have one intake port that turns into the cylinder wall (dashed line, bad) and one that turns into the center of the cylinder (solid line, good).

# GRAPHICAL ANALYSIS

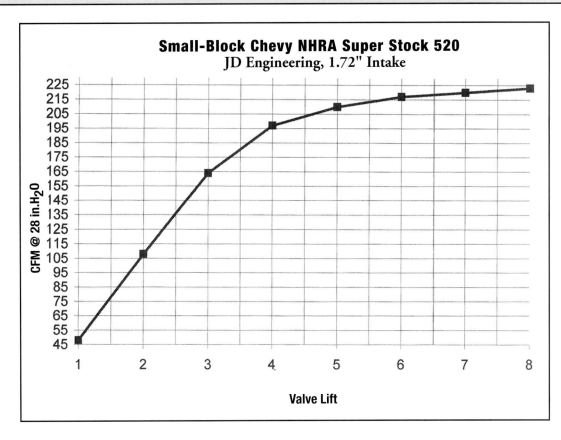

The Chevy small-block 520 cast iron casting has an ugly combustion chamber but check out the flow graph for a legal NHRA SuperStock application using a 1.72" intake valve. The rules are very restrictive on exactly what you can get away with in working with SuperStock cylinder heads.

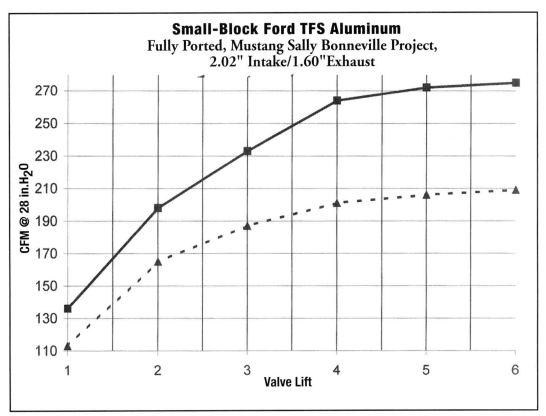

This graph shows how you can get airflow supplied to a 292 cid small-block Ford. Today there are lots of choices for aluminum heads for Fords. Note the valve sizes taken from the Chevy playbook.

# Engine Airflow

These Suzuki Hayabusa engines are absolutely amazing pieces. The graph shows the flow for the fully ported four-valve Hayabusa head. When turbocharged these little falcons can easily make over 500 hp and live! That is quite a lot on one tire patch!

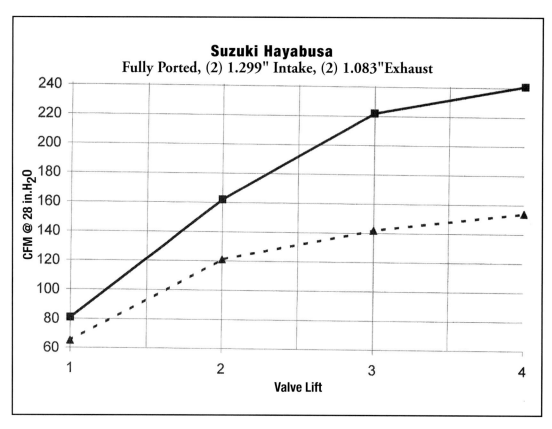

This engine has great performance potential for those willing to look outside the domestic V8 box. Stock flow vs. fully ported flow is shown. Typical of four-valve breathing. The engines are typically only 4L (244 cid) to about 4.6L (281 cid). Some are 5.7L (348 cid). All are EFI units. Provides possibilities for those with imagination.

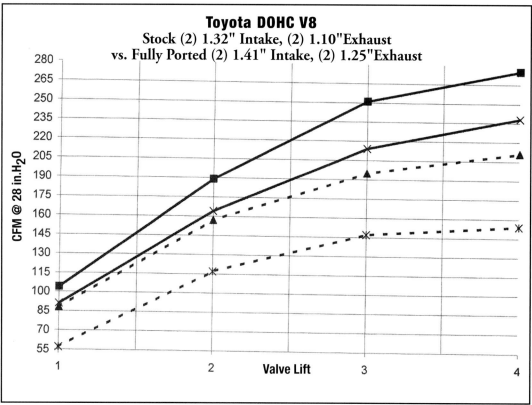

# Chapter 13
# Computer Simulation Programs

The Engine Expert program has been around for a number of years and has proven to be reliable. This shows many of the variables that can be input to evaluate a potential engine design. Photo courtesy Alan Lockheed.

*Do not say a little in many words but a great deal in a few.* —Pythagoras

There are many computer simulation programs that are in use in today's aftermarket industry and at the OEM levels. They range from the simplistic to the very complex and the pricing varies accordingly.

The OEMs typically use very expensive computer programs including CFD (computational fluid dynamics) and some of them are priced at upwards of hundreds of thousands of dollars per site. As CFD grows in capability and computers gain in performance and capacity perhaps one of these days the pricing of those type programs will decrease enough to be affordable by some regular gearheads.

Of the many engine simulations that can be used on personal computers, there are some that are very handy to use. Alan Lockheed's Engine Expert is good to use and has been around a long time and has proven to be quite reliable. Also handy to use is a Performance Trends package called Engine Analyzer or Engine Analyzer Pro with the Pro version having more features. Another that is easy to use and has active graphics is Dynomation that is sold by ProSim of CompCams. The Dynomation software has a great graphics display of intake and exhaust pressures as well as pressure in the cylinder and helps you to see and realize the importance of valve event timing and the effects of intake and exhaust changes. Patrick Hale's Racing Systems Analysis software for engines is called EnginePro and he has just finished a booklet that describes how he developed the software for his performance programs that he has been doing since 1986. Recently Pat sold his software packages to Don Terrill of Speed Talk, so it is available from there. Also available are software engine simulation and performance prediction programs from MaxRace and Pipe Max Software by Larry Meaux down in the swamps of Louisiana. One of the common bonds to all the software that is listed is their dependence on how the airflow capacity of the engine predicts a performance package or combination.

For all the computer operated simulation software programs listed, you need to have very good airflow data to feed into the simulation for it to give accurate and sensible results. If you inflate or decrease the numbers required for airflow of the cylinder heads and manifolds, you will get incorrect information as a result.

Most computer simulation programs are typically somewhere within +/- 2% to +/- 10% in predicting power accurately. It all depends on the quality of the input data. Theoretical horsepower will only win theoretical or virtual races, so at some point or another you will probably gravitate toward real parts and pieces to build something to race or at least see if it might run. Do your homework well and the end results will be worthwhile. Hey, if it was easy, everybody would be doing it!

# ENGINE AIRFLOW

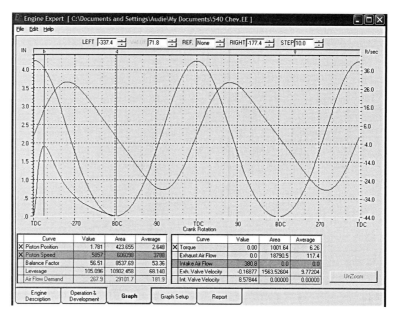

The Engine Expert program also has a graphical output that helps in the analysis of an engine design such as this piston position and crank position with a known rod length. Photo courtesy Alan Lockheed.

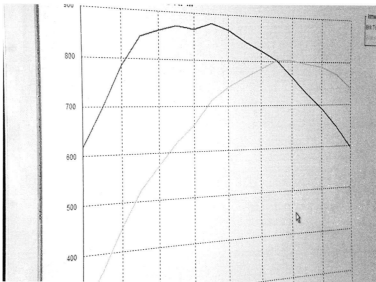

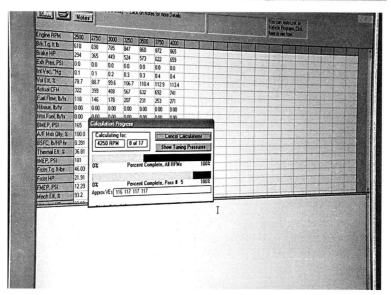

These screen images show one of the many capabilities of the simulation program from Performance Trends. Sometimes it is fun to build a dream engine and sort it out with a computer, but eventually you have to pick up hardware and begin for real. Photo courtesy 10 Litre Performance.

## IMAGINING ENGINE AIRFLOW

The following imagination exercise is presented as a way for you to form a mental picture of air and fuel as it flows through the engine. Admittedly, it is somewhat unusual to end a technoid book in this way, but the imagination is a wonderful tool to apply when thinking about airflow and engines as long as it is grounded in the basics. So let's get small and take a closer look at what happens just inside the intake port on an observation trip into the cylinder head.

Imagine that the port is a little over two inches high and we are only about one quarter of an inch tall, standing inside it, with the amazing capability to be firmly attached to the port walls wherever we stand. The engine starts and we immediately feel the vibration, a combination of the movement of the flow and the banging of the valves as they crash into the seats and that force is transferred to the cylinder head we are standing in. The engine is operating at 6000 rpm and it seems like being in the middle of a tornado or something. We feel the motion of the air trying to pull us into the long pathway toward the intake valve but we are securely positioned so that we can observe the air and fuel flow in the port of a running engine. Man is it noisy! Mixtures of air and fuel pressure waves zip past us at about 180 miles per hour on their way into the cylinder. As we look down the port toward the intake valve, we can see the blurred motion of the valve as it moves up and down at 50 times per second. The fog of the fuel suspended in and traveling with the air is awesome to view. The waves pulse to and fro and it is like being in a wind tunnel and seeing the smoke generators around a vehicle but this is far more exciting. This is the essence of fogged fuel that becomes power when it gets to the combustion chamber. For gearheads, this is better than any amusement ride we have ever experienced.

As we move toward the short side radius we can see that there are irregular ripples and changing turbulence of the fuel and air mix. Now it is much easier to understand what I was saying in the book about flow separation and attached flow. Wow! Look at how some of the fuel just sort of splatters against the backside of the valve as the valve goes toward the valve seat. We can see the depth of the fuel, called *wall flow*, is up to the top of our ankles as some of the slowly moving liquid flow is caught in the boundary layer flow. See how the valve has a bouncing motion to it as it hits the seat instead of closing just once? When the valve opens, the fuel splatter becomes part of the next batch of flow going into the combustion chamber as it is pulled off the edges of the valve. The valve opens at 2.6 times our height. Good thing that we are secure and safe in all this flow or we would end up being products of this engine's combustion and then get tossed out in the exhaust gases.

Wouldn't it be neat to be in the middle of the combustion chamber to see what goes on in there? As we get closer and closer to the edge of the valve seat area, we can feel the pulsations of the flow are timed with the motion of the valve and we begin to experience the increase of local temperature. Oddly enough we can plainly see that all portions around the valve are not flowing in the same fashion. Some of the heat from each combustion event is transferred to the space around us and begins to stabilize at about 100°F. Not uncomfortable, but not nearly as cool as it was at the entry to the port where it was only about 60°F.

Now, with the snap of our fingers, we are back outside the engine. Was that an amazing trip or what? Hopefully this visualization exercise will help you understand the role air and fuel flow play in creating horsepower.

# HANDY FORMULAE & CHARTS FOR GEARHEADS

Here in one place is an easy-to-find reference for many very handy formulae that are great to apply when you know where the numbers come from.

Note that in some of these equations torque can be listed in pounds-feet (lbs-ft) or as required in pound-inches (lb-in). The way to get to pound-inches is very easy to remember as 1 lb-ft = 12 lb-in.

## Horsepower
$hp = T \times rpm \div 5252$
where hp = horsepower, T = torque, lbs-ft, rpm = revolutions per minute

## Torque
$T = hp \times 5252 \div rpm$
where hp = horsepower, T = torque, lbs-ft, rpm = revolutions per minute

## Engine Rpm
$rpm = hp \times 5252 \div T$
where hp = horsepower, T = torque, lbs-ft, rpm = revolutions per minute

## Ram Air Pressure
$R_{air} = (mph)^2 \div 56{,}725$
where $R_{air}$ = air pressure in psi, mph is vehicle speed in miles per hour. Reciprocal of 56,725 is 0.0000176. This equation would be used to estimate the pressure available inside the air scoop or vent facing the direction of travel (in non-disturbed "clean" air). For reference, 0.1 psi = 2.7689886 in.$H_2O$. Otherwise, look at the appendix for equivalents or other units.

## Horsepower Required to Overcome Aerodymanic Air Resistance
$h_{pa} = (A \times C_d \times V^3) \div 146{,}600$
where $h_{pa}$ = horsepower requirement aerodynamic, A = frontal area of vehicle in square feet, $C_d$ = coefficient of drag, V = vehicle speed in miles per hour

## Horsepower Required to Overcome Rolling Resistance
$hpr_f = [(Rr_f \times V) \div 375]$
where $hpr_f$ = horsepower required to overcome rolling resistance, $Rr_f$ = rolling resistance force in lbs, V = vehicle velocity in mph

## Rolling Resistance Calculation
$Rr_f = Cr_f \times W$
Where $Rr_f$ = rolling resistance force (pounds), $Cr_f$ = coefficient of rolling friction (typical auto tires at normal inflation = 0.015 or 0.016), W = vehicle weight (lbs.)

## Horsepower Based on Quarter-Mile Speed
$hp = (0.00426 \times mph)^3 \times Veh\ Wt$
where hp = horsepower at the flywheel, mph = vehicle terminal speed in quarter-mile, Veh Wt = vehicle weight in lbs.

## Elapsed Time Based on Quarter-Mile Speed
$ET_{est} = 5.5 \times (235 \div V_v) \times \sqrt[3]{[((V_v - 100 \div 400)) + 1]}$
where $ET_{est}$ = estimated elapsed time in seconds, $V_v$ = quarter-mile trap speed of vehicle in mph, note there is a cube root function in this equation. Not as dependable as formula listed above.

## Tractive Effort
$TE = (T \times G_r \times e \times 12) \div r$
where TE = the total force available to propel a vehicle, T = observed engine torque in lbs-ft. at flywheel, $G_r$ = overall gear ratio, e = mechanical efficiency of drivetrain (0.85 to 0.95, 0.90 is most common), r = rolling radius of drive axle tire (inches)

## Drawbar Pull
$DP = ((T \times G_r \times e) \div r) - Rr_f$
where DP = drawbar pull in pounds force, T = engine torque (lb-in.), $G_r$ = overall gear ratio, e = efficiency of drivetrain (0.85 to 0.95, 0.90 is common), r = rolling radius of drive axle tire (inches), $Rr_f$ = rolling resistance force (pounds)

## Engine Torque Required to Slip Tires
$S_t = (W_{vda} \times \mu \times r) \div (G_r \times e)$
where $S_t$ = engine torque required to slip tires (lbs-in., divide by 12 to yield an answer in lbs-ft.), $W_{vda}$ = weight over drive axle (pounds), $\mu$ = Greek letter mu (coefficient of friction), r = rolling radius of drive axle tire (inches), $G_r$ = overall gear ratio, e = efficiency of drivetrain (0.85 to 0.95, 0.90 is common)

## Miles per Hour
$mph = (D_t \times rpm) \div (336 \times G_r)$
where mph = miles per hour, $D_t$ = diameter of drive tire in inches, rpm = engine revolutions per minute, $G_r$ = gear ratio

## Tire Size
$D_t = (336 \times G_r \times mph) \div rpm$
where $D_t$ = diameter of drive tire in inches, $G_r$ = gear ratio, mph = miles per hour, rpm = engine revolutions per minute. To find rpm from the gear ratio and tire size:
$rpm = (336 \times G_r \times mph) \div D_t$
where rpm = engine revolutions per minute, $G_r$ = gear ratio, mph = miles per hour, $D_t$ = diameter of drive tire in inches

## Gear Ratio
$G_r = (D_t \times rpm) \div (336 \times mph)$
where $G_r$ = gear ratio, $D_t$ = diameter of drive tire in inches, rpm = engine revolutions per minute, mph = miles per hour

## Coefficient of Drag (Coast-Down Method)

**Note:** You are responsible for your own actions and the author or publishers do not recommend that you exceed the speed limit. The coast-down test is best performed on a deserted stretch of flat road surface, at your own risk.

$C_d = K \times (1/V_f - 1/V_i) \div ET$

where $C_d$ = coefficient of drag, $K = (w \div g) \div 0.5 \times \rho_s \times F_a$, $V_f$ = velocity final, $V_i$ = velocity initial, W = total vehicle weight lbs, g = 32.17, $\rho_s$ = density in slugs, $F_a$ = Frontal area in square feet, ET = stopwatch elapsed time in seconds between initial and final velocities. Refer to air density chart on page 148.

Another method is to take the test at high speed (such as 60 mph and let the vehicle coast to 55 mph). Then average the two by adding 60 + 55 and divide by 2. The average speed divided by time will give you an average deceleration rate.

Doing the same thing for lower speeds over the same test real estate such as 40 mph to 35 mph will give you the average speed and becomes the average decel rate when divided by time. Choose a clear day as the results are greatly affected by wind and rain. **Hint:** The longer it takes to slow down from one speed to a lower speed shows the vehicle is "slicker" and also has less rolling resistance.

## Estimated Horsepower from Airflow

$hp_{est} = cfm_{28} \times 0.257 \times N_{cyl}$

where $hp_{est}$ = estimated horsepower, $cfm_{28}$ = single port (including manifold and throttle body or carburetor) airflow at 28 in.$H_2O$, $N_{cyl}$ = number of cylinders in engine

## Estimated Power Related to Fuel Flow

hp x bsfc = Fuel Flow

hp = horsepower, BSFC = brake specific fuel consumption (use 0.5 for gasoline, 1.0 to 1.04 for methanol, ~0.6 to 0.65 for E85), Fuel Flow = lbs/hr

## Horsepower Increase from Supercharging

$hp_{bln} = [(b_{psi} + A_{psi}) \times hp_{na}] \div A_{psi}$

where $hp_{bln}$ = estimated blown hp, $b_{psi}$ = boost in psi, $A_{psi}$ = atmospheric pressure in psi, $hp_{na}$ = hp naturally aspirated

## Effective Compression Ratio from Supercharging

$CR_{bln} = [(b_{psi} + A_{psi}) \times CR] \div A_{psi}$

where $CR_{bln}$ = Compression ratio blown, $b_{psi}$ = boost in psi, $A_{psi}$ = atmospheric pressure in psi, CR = static compression ratio

## Piston Speed

$P_s = (S \times rpm) \div 6$

where $P_s$ = piston speed in feet per minute, S = stroke in inches, rpm = engine revs per minute. If you want a piston speed in ft/sec, divide answer by 60

## Port Velocity

$P_{vel} = (P_s \div 60) \times (B^2 \div A_p)$

where $P_{vel}$ = port velocity in feet per second, $P_s$ = piston speed in feet per minute, B = bore diameter in inches (squared here), $A_p$ = area of port in square inches

## Valve Curtain Area (Simple)

$A_c = \pi \times D_v \times L_v$

where $A_c$ = curtain area, $D_v$ = diameter of valve, $L_v$ = lift of valve

## Valve Curtain Area (Complex)

$A_c = \pi \times (D_v - W_s) \times 2 \times \cos(\alpha)$

where $A_c$ = curtain area, $D_v$ = diameter of valve, $W_s$ = width of seat (inches), $\cos(\alpha)$ = cosine of the angle of the seat. See chart on page 147.

## Valve Area

$A_v = (D^2 \times 0.7854) - (d^2 \times 0.7854)$

where $A_v$ = effective area of the valve in square inches ($in^2$), D = major diameter of valve in inches, $d^2$ = diameter of valve stem in inches

## Engine Displacement

$D_e = B^2 \times S \times 0.7854 \times N_{cyl}$

where $D_e$ = engine displacement in cubic inches, B = diameter of bore in inches, S = stroke in inches, $N_{cyl}$ = number of cylinders.

## Power Density

$P_{den} = hp/in^3$

where $P_{den}$ = power density, hp = horsepower, $in^3$ = displacement of engine, cubic inches

## Power Referenced to Piston Area

$P_{pa} = CB_{hp} \div A_p$

where $P_{pa}$ = power per piston area in hp ÷ $in^2$, $CB_{hp}$ = corrected brake horsepower, $A_p$ = area of piston in square inches ($in^2$)

## Torque per Cubic Inch

$T \div D_e$

where T = lb-ft, $D_e$ = displacement of engine in cubic inches

## Indicated Horsepower

$I_{hp} = PLANK \div 33,000$

where $I_{hp}$ = indicated horsepower, P = mean effective pressure in psi, L = length of stroke in feet, A = cylinder area in square inches, N = number of power strokes per minute which is rpm ÷ 2, K = number of cylinders. Remember here that IMEP – FMEP = BMEP and $I_{hp} = F_{hp} + B_{hp}$

# Handy Formulae & Charts for Gearheads

## BMEP
$BMEP = 150.8 \times (T \div D_e)$
where BMEP = brake mean effective pressure in pounds per square inch, T = torque in lb-ft, $D_e$ = displacement of engine in cubic inches

## Comparison of Airflow at Different Test Pressures
$Q = \sqrt{(P_1 \div P_2)} \times q$
where Q = resultant flow, $P_1$, $P_2$ are pressures for test and comparison, q = original flow number

## Nozzle or Injector Fuel Flow at Different Pressures
$F_{new} = \sqrt{(P_{new} \div P_{old})} \times F_{old}$
where $F_{new}$ = new flow rate (lbs. or gal. per time), $P_{new}$ = new pressure (psi), $P_{old}$ = old pressure (psi), $F_{old}$ = old flow rate (lbs. or gal per time)

## Relative Air Density
$RAD = ((P_b - V_p) \div 29.92) \times (520 \div T_{air})$
where RAD = relative air density, $P_b$ = local barometric pressure, in.Hg, Vp = vapor pressure, in.Hg, $T_{air}$ = °F + 460. At 29.92 in.Hg, 60°F, dry air (0), the RAD = 1.00

## Air Density Percentage
$100 \times (P - V_p \div 29.92) \times (520 \div T_{air})$
where $P_b$ = local barometric pressure, in.Hg, $V_p$ = vapor pressure, in.Hg, $T_{air}$ = °F + 460

## Mass Flow Through Sharp-Edged Orifice
$Q_m = 0.09970190 \times C \times Y \times d^2 \times \sqrt{(h_w \times \rho_f) \div (1 - \beta^4)}$
where $Q_m$ = mass rate of flow in lbs/sec, C = coefficient of discharge, Y = expansion factor, d = diameter of orifice, inches, $h_w$ = differential pressure, in.$H_2O$, $\rho_f$ = Greek letter rho, density of flowing fluid, $\beta$ = diameter ratio

## Volume Flow Through Edged Orifice
$Q = A (C_d K p_1)$
where $C_d$ = Coefficient of orifice, K = 4005 (this constant is for a sea level reference where the air density is typically 0.075 lbs/ft$^3$, $P_1$ = pressure differential in inches of $H_2O$.
$q = 411 \times (d^2)(1.02 \times 10^{-3})(T_p)$
here q = flow in cfm (cubic feet per minute), d = diameter of orifice in inches, Tp = test pressure, in.$H_2O$.

Many assumptions are made with this simple equation, one being that the fluid is air, another being that the $C_d$ is held constant at 0.60, so round orifices in thin material are allowed for in this equation.

Just for fun: 1.875" orifice, 28 in.$H_2O$ test pressure would yield about 244.19 cfm while the equation below would yield 247.23 cfm. The "real" answer is about 249 to 250 cfm. You should be able to work on these things and decide what you want to use…

## Cubic Feet per Minute
$cfm = 13.29 \times d^2 \times \sqrt{P_t}$
where cfm = cubic feet per minute, d = diameter of orifice in inches, $P_t$ = test pressure inches of water

## Airflow Rate (Metric)
This one is in metric units for those of you that think those numbers are more convenient to deal with:
$Q = 0.1864 C_d d^2 \sqrt{h T_a \div p_a}$
where Q = Airflow in cubic meters per second, $C_d$ = Coefficient of Discharge, $d^2$ = diameter of orifice (mm), squared, h = head (mm $H_2O$), $T_a$ = Temperature absolute (degrees C + 273), $p_a$ = pressure absolute

## Flow Bench Fundamentals and How to Calculate the Effective Flow Area (EFA) Applying Fleigner's Equation
$acfm = [(29.92 \div P_b) \times ((T + 460) \div 520)] \times F_b$ Flow
where acfm = actual cubic feet per minute, $P_b$ = local barometer in.Hg, T = °F, $F_b$ Flow = flow number from flow bench (cubic feet per minute)

$\rho = 1.325 \times [P_b \div (T + 460)]$
where $\rho$ = Greek letter rho, local air density lbs/ft$^3$, $P_b$ = local barometer, in.Hg, T = °F

$Q_m = (\rho \times acfm) \div 60$
where $Q_m$ = Mass Flow lbs/sec, $\rho$ = local air density lbs/ft$^3$, acfm = actual cubic feet per minute

$P_r = P_b \div (P_b - T_p)$
where $P_r$ = pressure ratio, $P_b$ = local barometer, in.Hg, $T_p$ = test pressure, psi

$M = \sqrt{[(P_r) \times 0.2857 - 1 \div 0.2]}$
where M = Mach number, $P_r$ = pressure ratio

$A_e = EFA = [(Q_m \div P_b) \div \sqrt{(T + 460)} \times (0.91886 \times M) \times (1 + 0.2 \times M^2)^{-3}]$
where $A_e$ = EFA = effective flow area, $Q_m$ = mass flow, lb/sec, $P_b$ = local barometric pressure, in.Hg, T = Temperature °F, M = Mach number

## Estimated Power of Supercharged Engines

This provides a rough estimation of power added relative to boost on supercharged engines. Somewhat simplistic, but at least it is quick:

$HP+ = \sqrt{[(B_{psi} + D) \div (B_{psi} - V)]}$

where HP+ = horsepower increase, $B_{psi}$ = local atmospheric barometric pressure expressed in psi (RAD is preferred, but more difficult to calculate, see listing note below), D = relative density index (boost pressure × efficiency), V = full throttle vacuum, normally aspirated (typically between 0.75 psi and 3 psi).

This is one way to estimate the volumetric efficiency of the engine package if it was naturally aspirated as a baseline. At WOT, most modified engines will attain about 0.8 in.Hg or approximately 10.88 in.$H_2O$ or about 0.393 psi.

**Example:** 10 psi of boost (estimated at 65% efficiency) is applied to an engine that pulls 2 psi of vacuum at WOT. With elevation at 3000 feet, local atmospheric pressure is 13.26 psi. D = 10 psi × 0.65 = 6.5.
$B_{psi} + D = 13.26 + 6.5 = 19.76$.
$B_{psi} - V = 13.26 - 2 = 11.26$. $19.76 \div 11.26 = 1.7549$.
HP+ = the square root of 1.7549, which is 1.325. So, adding 10 psi in this case adds about 32.5% power.
$RAD = [(B_p - V_p \div 29.92) \times (520 \div T_{air} + 460)] \times 100$
where RAD = relative air density in percent, $B_p$ = barometric pressure (in.Hg), $V_p$ = vapor pressure (in.Hg), $T_{air}$ = air temperature, °F.

### Cosines of Angles
(for valves and seats)

| Angle(α) | Cosine | Angle(α) | Cosine |
|---|---|---|---|
| 30° | 0.866025 | 47° | 0.681998 |
| 35° | 0.819152 | 50° | 0.642788 |
| 37° | 0.798636 | 52° | 0.615661 |
| 40° | 0.766044 | 55° | 0.573576 |
| 42° | 0.743145 | 57° | 0.544639 |
| 45° | 0.707107 | 60° | 0.500000 |

### Coefficients of Friction/Adhesion
(μ) (between rubber tire and material)
To be used with Equation $F = \mu \times N$

| Material surface | Coefficient of friction (μ) |
|---|---|
| Concrete | 1.02 |
| Asphalt | 0.72 |
| Rubber | 1.16 |
| Salt (hard packed) | 0.4 to 0.5 |
| Snow (hard packed) | 0.25 to 0.35 |
| Sand (hard packed) | 0.3 to 0.5 |
| Clay (hard packed) | 0.52 to 0.65 |
| Gravel | 0.6 to 0.7 |
| Ice | 0.15 to 0.2 |

### Streamlined Orifice Flow Characteristics

Based on Q = AV with streamlined orifice of 1 $in^2$ (0.0069444 $ft^2$) with a $C_d = 1.00$

| Test press In.$H_2O$ | CFM per $in^2$ | Velocity (ft/sec) |
|---|---|---|
| 15 | 106.81 | 256.34 |
| 20 | 123.33 | 295.99 |
| 20.4 | 124.56 | 298.94 |
| 25 | 137.89 | 330.93 |
| 28 | 145.93 | 350.23 |
| 30 | 151.05 | 362.52 |
| 35 | 163.15 | 391.57 |
| 40 | 174.42 | 418.60 |
| 45 | 184.99 | 443.99 |
| 50 | 195.00 | 468.01 |
| 55 | 204.52 | 490.85 |
| 60 | 213.62 | 512.68 |
| 65 | 222.34 | 533.61 |
| 70 | 230.73 | 553.76 |
| 75 | 238.83 | 573.19 |
| 80 | 246.66 | 591.99 |
| 85 | 254.25 | 610.21* |
| 90 | 261.62 | 627.90* |
| 95 | 268.79 | 645.10* |
| 100 | 275.77 | 661.85* |

*Above Mach 0.55 past this point

# Zicht Calibration Sheet for Gasoline Fuel

| Jet# | Size | Cal# | Jet# | Size | Cal# |
|---|---|---|---|---|---|
| 50 | 0.048 | 3.394 | 80 | 0.087 | 7.781 |
| 51 | 0.049 | 3.499 | 81 | 0.089 | 8.009 |
| 52 | 0.050 | 3.605 | 82 | 0.090 | 8.149 |
| 53 | 0.051 | 3.712 | 83 | 0.093 | 8.472 |
| 54 | 0.052 | 3.821 | 84 | 0.094 | 8.615 |
| 55 | 0.053 | 3.930 | 85 | 0.095 | 8.758 |
| 56 | 0.054 | 4.040 | 86 | 0.096 | 8.902 |
| 57 | 0.055 | 4.152 | 87 | 0.098 | 9.140 |
| 58 | 0.056 | 4.264 | 88 | 0.100 | 9.380 |
| 59 | 0.057 | 4.378 | 89 | 0.101 | 9.528 |
| 60 | 0.058 | 4.492 | 90 | 0.102 | 9.676 |
| 61 | 0.059 | 4.607 | 91 | 0.104 | 9.920 |
| 62 | 0.060 | 4.724 | 92 | 0.106 | 10.160 |
| 63 | 0.061 | 4.841 | 93 | 0.108 | 10.410 |
| 64 | 0.062 | 4.959 | 94 | 0.111 | 10.760 |
| 65 | 0.063 | 5.079 | 95 | 0.112 | 10.910 |
| 66 | 0.064 | 5.199 | 96 | 0.113 | 11.070 |
| 67 | 0.065 | 5.320 | 97 | 0.116 | 11.420 |
| 68 | 0.066 | 5.442 | 98 | 0.117 | 11.580 |
| 69 | 0.067 | 5.565 | 99 | 0.118 | 11.740 |
| 70 | 0.071 | 5.940 | 100 | 0.120 | 11.990 |
| 71 | 0.075 | 6.319 | 101 | 0.121 | 12.160 |
| 72 | 0.076 | 6.448 | 102 | 0.122 | 12.320 |
| 73 | 0.077 | 6.578 | 103 | 0.123 | 12.480 |
| 74 | 0.078 | 6.709 | 104 | 0.124 | 12.640 |
| 75 | 0.079 | 6.841 | 105 | 0.125 | 12.800 |
| 76 | 0.080 | 6.974 | 106 | 0.126 | 12.970 |
| 77 | 0.083 | 7.283 | 107 | 0.127 | 13.130 |
| 78 | 0.084 | 7.418 | 108 | 0.128 | 13.300 |
| 79 | 0.086 | 7.643 | 109 | 0.129 | 13.460 |

# Critical Temperatures for Various Fuels

| Fuel | Auto Ignition | Boiling Point |
|---|---|---|
| Gasoline | 536°F | 100–400°F* |
| Iso-octane | 837°F | 244°F |
| Methanol | 725°F | 149°F |
| Ethanol | 689°F | 172°F |
| E85 | ~55 –600°F | 96–170°F* |
| Nitromethane | 785°F | 214°F |
| Toluene | 849°F | 321°F |
| Benzene | 1040°F | 176°F |
| Acetone | 869°F | 134°F |
| Ethylene | 914°F | –155°F |
| Propane | 842°F | –44°F |
| Methane | 1076°F | –259°F |
| Xylene | 867°F | 281°F |

*Fuels with two temperature references denote light ends at first temperature and heavy ends with second temperature. Sometimes this is called point 1 and point 2 on the distillation curve for the fuel.

# Air Density at Sea Level

| Temp °F | Air Density $\rho$ (slugs/ft$^3$) | Specific Weight (lb/ft$^3$) |
|---|---|---|
| 0 | 0.002683 | 0.08633 |
| 10 | 0.002626 | 0.08449 |
| 20 | 0.002571 | 0.08273 |
| 30 | 0.002519 | 0.08104 |
| 40 | 0.002469 | 0.07942 |
| 50 | 0.002420 | 0.07786 |
| 60 | 0.002373 | 0.07636 |
| 70 | 0.002329 | 0.07492 |
| 80 | 0.002286 | 0.07353 |
| 90 | 0.002244 | 0.07219 |
| 100 | 0.002204 | 0.07090 |
| 110 | 0.002168 | 0.06968 |
| 120 | 0.002128 | 0.06846 |

## UNIT CONVERSIONS & EQUIVALENTS

| | | | |
|---|---|---|---|
| 1 cfm | 0.02831685 m³/min | 28.31685 L / min | 0.4719474 L /sec |
| 1 gpm | 3.785412 L/min | 0.003785412 m³/min | 63.0902 cc/sec |
| | | | |
| 1 hp | 42.44 Btu/min | 0.746 kilowatts | 178.3 calories/sec |
| 1 hp | 2546.7 Btu/hr | 745.6999 watts | 550 ft-lbs/sec |
| | | | |
| 1 mph | 1.47 ft/sec | 1.61 km/hr | 0.447 m/sec |
| 1 mph | 0.8689762 knot/hr | 1.467 ft/sec | 44.7 cm/sec |
| 1 knot | 1.150779 mph | 1.68781 ft/sec | 1.852 km/hr |
| 1 km/hr | 0.6213712 MPH | 0.5399568 knot/hr | 0.9113444 ft/sec |
| 60 mph | 88 ft/sec | 96.56 km/hr | 26.82 m/sec |
| 1 Mach | 741.4549 mph | 1087.467 ft/sec | 331.46 m/sec |
| | | | |
| 1 foot | 12.00 in. | 30.48 cm | 304.8 mm |
| 1 inch | 25.4 mm | 2.54 cm | 0.0254 m |
| 1 meter | 39.36996 inches | 1.09361 yards | 0.0006213712 mile |
| 1 mile | 1609.344 meters | 1760 yards | 5280 feet |
| 1/4 mile | 1320 feet | 440 yards | 0.4025 km |
| 1 mm | 0.03936996 inch | 0.001 meter | |
| 1 micron | 0.00003936996 inch | | |
| | | | |
| 1 psi | 2.036021 in.Hg | 6.894757 kpa | 27.689885 in.H₂O |
| 1 inch Hg | 13.6 in.H₂O | 3.386388 kpa | 0.4911541 psi |
| 1 bar | 29.52999 in.Hg | 401.60788 in.H₂O | 100 kpa |
| | | | |
| 1 pound | 0.4536 Kg | 16 ounces | 453.6 grams |
| 1 Kg | 2.2046 pounds | 1000 grams | 35.274 ounces |
| 1 lb-ft | 12 lb-in | 1.35882 N-m | 0.1382552 kg-m |
| | | | |
| 1 deg. F | −17.222 deg. C | 255.9278 deg. Kelvin | 460.67 Rankine |
| 32 deg. F | 0 deg. C | 273.15 deg. Kelvin | 491.67 Rankine |
| 60 deg. F | 15.555 deg. F | 288.7056 deg. Kelvin | 519.67 Rankine |
| 212 deg. F | 100 deg. C | 373.15 deg. Kelvin | 671.67 Rankine |
| 1 deg. C | 33.8 deg. F | 274.15 deg. Kelvin | 493.47 Rankine |

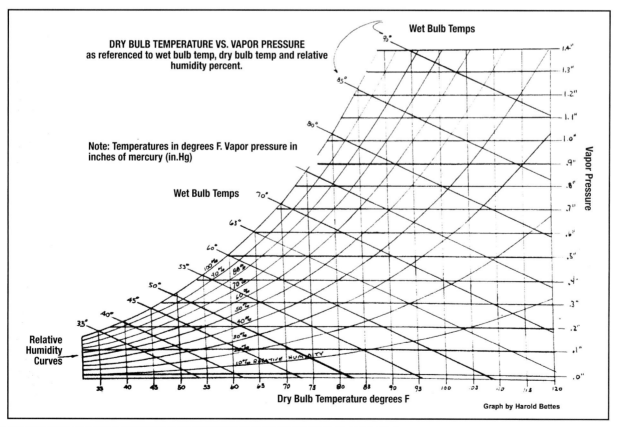

**DRY BULB TEMPERATURE VS. VAPOR PRESSURE** as referenced to wet bulb temp, dry bulb temp and relative humidity percent.

Note: Temperatures in degrees F. Vapor pressure in inches of mercury (in.Hg)

Graph by Harold Bettes

This chart can be a very handy trackside tuning reference.

# RESOURCES

**10 Litre Performance**
Specialty cylinder heads for H-D, automotive and marine, flow testing
11915 119th Pl.
Northglenn, CO 80233
(303) 280-9950
www.10litre.com

**Alumilite Corporation**
Quick set mold rubber for port molding
315 E. North St.
Kalamazoo, MI 49007
(800) 447-9344
www.alumilite.com

**Audie Technology, Inc.**
Flow bench instrumentation and controls (for DIY flow benches)
23 N. Trooper Rd.
Trooper, PA 19403
(610) 630-5895
www.audietech.com

**Bill Jones**
Porting, flow testing & procedures
3294 W. Meadow Wood Way
Taylorsville, UT 84118-2815
(801) 969-3807
http://www.ryanbrownracing.com/Bill_Jones_Photo_Gallery.html

**Calspec CNC**
Flow bench accessories and specialty measuring tools for shops
Woodland Park, CO 80863
(719) 439-6406
www.calspeccnc.com

**CFM Performance Carburetors**
3337 Yost Rd.
Litchfield, OH 44253
(330) 723-5683
www.cfmperfcarbs.com

**Conley Precision Engines**
Miniature running engines and vehicles
825 Duane St.
Glen Ellyn, IL 60137
(630) 858-3160
www.conleyprecision.com

**Dow Corning Corporation**
Silastic RTV S-5370 Kit for port mold making
PO Box 994
Midland, MI 48686
(989) 496-7875
www.dowcorning.com

**DRPro**
Drag Racing Pro computer software and consulting
**Patrick Hale**
12821 N. 18th Pl.
Phoenix, AZ 85022-5736
(602) 992-2586
www.DragRacingPro.com

**Dwyer Instruments, Inc.**
Manometers and various accessories for flow measurement
PO Box 373
Michigan City, IN 46361
(219) 879-8000
www.dwyer-inst.com

**Engine Expert**
Engine performance simulation software
PO Box 5142
Golden, CO 80401-0501
(303) 238-2414
alockheed@netzero.net

**Fleming & Associates, Inc.**
Aerospace flow benches and airflow calibration equipment
1060 North Capitol Ave. Suite E-100
Indianapolis, IN 46204
(317) 631-4605

**Flow Data**
Custom Flow Benches
621 S. East St.
Anaheim, CA 92805
(714) 817-0567
www.flowdata.net

**Flow Performance, LLC**
Instrumentation and flow computers
Novato, CA 94948
(415) 892-8158
www.flowperformance.com
(479) 841-8244

**Flow Systems**
Calibration services and flow benches
220 Bunyan Ave.
Berthoud, CO 80513
(970) 532-0617
www.flowsystemsinc.com

**Jamison Equipment Company**
Exclusive distributor of Saenz flow benches and accessories
1908 11th St.
Emmetsburg, IA 50536
(800) 841-5405
www.jamisonequipment.com

**JD Engineering**
Cylinder head work, flow bench testing
114 East Bend Ct., Unit 2
Mooresville, NC 28117
(704) 658-6951

**Magnafuel Racing Fuel Systems**
615 Wooten Road, Suite 120
Colorado Springs, CO 80915
(719) 532-1897
www.magnafuel.com

**Meaux Racing Heads**
MaxRace Software
Engine and performance simulation software, including PipeMax for intake and exhaust tuning
9827 Louisiana Hwy. 343
Abbeville, LA 70510
(337) 652-6220
www.maxracesoftware.com

**Meriam Process Technologies**
Laminar flow elements, manometers, accessories for flow measurement
10960 Madison Ave.
Cleveland, OH 44102
(216) 281-1100
www.meriam.com

**Midwest Cylinder Head**
Cylinder head repair, including welding cast iron ports and chambers)
1700 West F Ave.
Nevada, Iowa 50201
(515) 382-2791

# RESOURCES

**Mondello Tech Center**
Head porting, flow bench accessories, wet flow bench
2470 Pomona Rd.
Crossville, TN 38571
(931) 459-2760
www.mondello.com

**Nat's Racing Engines**
Racing engines, machine work, flow testing
702 Warren Ave.
Swansea, MA 02777
(508) 336-4142

**Omega Engineering Inc.**
Instrumentation for flow measurement, pressure and flow calibration services
One Omega Dr.
PO Box 4047
Stamford, CT 06907
(203) 359-1660
www.omega.com

**Performance Trends**
Engine and performance simulation software
PO Box 503164
Livonia, MI 48153
(248) 473-9230
www.performancetrends.com

**Perma-Flex Mold Company, Inc.**
Mold maker's silicone rubber Blu-Sil and others
1919 E. Livingston Ave.
Columbus, OH 43209
(800) 736-6653
www.perma-flex.com

**Precision Measurement Supply**
Tooling and accessories
7050 Snowflake Dr.
San Antonio, TX 78228
(210) 681-2405
www.precisionmeasure.com

**QuarterJr.com**
Engine simulation software
PO Box 527
Oshtemo, MI 49077
www.quarterjr.com

**Roadrunner Engineering**
Flathead specialty components, intake manifolds, superchargers, flow testing
PO Box 53296
Albuquerque, NM 87153
www.roadrunnerengineering.com

**Sean Murphy Induction**
17671 Metzler Ln. Unit A-7
Huntington Beach, CA 92647
(714) 843-9169
www.smicarburetor.com

**Superior Signal Company, Inc.**
Smoke candles for flow testing
PO Box 96
Spotswood, NJ 08884
(800) 945-8378
www.superiorsignal.com

**Thorpe Engine Development**
Flow balls, flow flags and flow bench development tools
3182 Crow Bench Rd.
Orofino, ID 83544
(208) 476-7572
www.thorpedev.com

**Virkler & Bartlett LLC**
Vintage racing engines and specialty components
1975 Slatesville Rd.
Chatham, VA 24531
(434) 432-4409
www.vbengines.com

# RECOMMENDED READING

There are a number of books, reports and technical papers that are absolutely necessary to have for reference or just for reliable reading about engines and airflow. However, view the accompanying list and consider building or adding to your own library, or it is also worthwhile to locate them in your public library for quick reference. They are all strongly recommended references.

## Books
*The Gas and Oil Engine*, Dugald Clerk, 1899, John Wiley and Sons

*The High Speed Internal Combustion Engine*, Sir Harry R. Ricardo, 1960, Blackie and Son Limited

*Applied Carburation and Petrol Injection*, Charles H. Fisher, 1951, Chapman and Hall, Ltd.

*Engineering Thermodynamics*, Short, Kent and Treat, 1953, Harper & Brothers

*Hydraulics and Fluid Mechanics*, Ranald V. Giles, 1956, Schaum Publishing

*Elementary Fluid Mechanics*, John K. Vennard, 1958, John Wiley and Sons

*Fluid Mechanics with Engineering Applications*, Robert L. Daugherty and Joseph B. Franzini, 1965, McGraw-Hill, ISBN 0-07-015427-9

*Fluid-Dynamic Drag*, Sighard F. Hoerner, 1965, Hoerner Fluid Dynamics

*Analytical Mechanics for Engineers—Dynamics*, Charles L. Best and William G. McLean, 1966, International Textbook Company

*Internal Combustion Engines in Theory and Practice*, Volumes I & II, C. F. Taylor, 1966, MIT Press, ISBN 0-262-70015-8

*Experimental Methods for Engineers*, J.P. Holman, 1966, McGraw-Hill, ISBN 0-07-029613-8

*Carburetors and Carburetion*, Walter B. Larew, 1967, Chilton Book Company

*Internal Combustion Engines and Air Pollution*, Edward F. Obert, 1973, Harper & Row, ISBN 0-352-04560-0

*Introduction to Fluid Mechanics*, Fox & McDonald, 1973, John Wiley and Sons, ISBN 0-471-01909-7

*Engineering Fluid Mechanics*, J.A. Fox, 1974, McGraw-Hill, ISBN 0-07-021750-5

*Fluid Mechanics*, Frank M. White, 1979, McGraw-Hill, ISBN 0-07-069667

*Internal Combustion Engines Applied Thermosciences*, Colin R. Ferguson, 1986, John Wiley and Sons, ISBN 0-471-88129-5

*Internal Combustion Engine Fundamentals*, John B. Heywood, 1988, McGraw-Hill, ISBN 0-07-028637-X

*Applied Combustion*, Eugene Keating, 1993, Marcel Dekker, Inc., ISBN 0-8247-8127-9

*Engine Testing Theory and Practice*, M. Plint and A. Martyr, 1995, Butterworth—Heinmann, ISBN 0-7506-1668-7

*The Metrology Handbook*, Jay L. Bucher, 2004, American Society for Quality, ISBN 978-0-87389-620-7

*Dyno Testing and Tuning*, Harold Bettes and Bill Hancock, 2008, SA 138, CarTech Books, ISBN 13: 978-1-932494-49-5

## National Advisory Committee for Aeronautics (NACA) Technical Papers
*Airflow Through Poppet Valves*, G. W. Lewis, E.M. Nutting, 1918, Report 24

*4th Annual Report of NACA*, NACA

*Intermittent-Flow Coefficients of a Poppet Valve*, C.D. Waldron, 1939, NACA Technical Note 701, NACA

## American Society of Mechanical Engineers (ASME) Technical Papers
*Pressure Drop Through Poppet Valves*, C. E. Lucke, 1905, Transactions of American Society of Mechanical Engineers, Vol. 27

*Short Pipe Manifold Design for Four-Stroke Engines*, Peter C. Vorum, 1980, Paper No. 80-DGP-6, ASME

*A Non-Linear Acoustic Model of Inlet and Exhaust Flow in Multi-Cylinder Internal Combustion Engines*, M. Chapman, J.M. Novak, R.A. Stein, 1983, Paper No. 83-WA/DSG-14

*Measurement of Fluid Flow in Pipes Using Orifice, Nozzle, and Venturi*, American Society of Mechanical Engineers, 1990, ASME Flow Standard, ASME

## Society of American Engineers (SAE) Technical Papers
*Airflow Through Intake Valves*, G.B. Wood, D.U. Hunter, E.S. Taylor, C. F. Taylor, 1942, SAE Journal (Transactions), Vol. 50, No. 6

*Aircraft—Engine Inlet and Exhaust Porting*, Vincent C. Young, 1944, SAE Journal (Transactions), Vol. 52, No. 5

*A Study of Engine Breathing Characteristics*, Carl H. Wolgemuth, Donald R. Olson, 1965, 650448, SAE

*Research and Development of High-Speed, High-Performance, Small Displacement Honda Engines*, Shizuo Yagi, Akira Ishizuya, Isao Fujii, 1970, 700122, SAE

*Design Refinement of Induction and Exhaust Systems Using Steady-State Flow Bench Techniques*, G. F. Leydorf Jr., R. G Minty, M. Fingeroot, 1972, 720214, SAE

*Measurement of Induction Gas Velocities in a Reciprocating Engine Cylinder*, M.J. Arnold, M.J. Tindal, T.J. Williams, 1972, 720115, SAE

*An Analytical Model for Exhaust System Design*, L.J. Eriksson, 1978, 780472, SAE

*Effect of Exhaust System Design on Engine Performance*, Tim G. Adams, 1980, 800319, SAE

*Airflow through Poppet Inlet Valves—Analysis of Static and Dynamic Flow Coefficients*, Itaru Fukutani, Eiichi Watanabe, 1982, 820154, SAE

*Characterization of Flow Produced by a High-Swirl Inlet Port*, Teoman Uzkan, Claus Borganakke, Thomas Morel, 1983, 830266, SAE

*Unsteady Flow Velocity Measurements Around an Intake Valve of a Reciprocating Engine*, Sherif H. Tahry, Bahram Khalighi, William R Kuziak, Jr., 1987, 870593, SAE

# RECOMMENDED READING

*A Transparent Engine for Flow and Combustion Visualization Studies,* Stephen C. Bates, 1988, 880520, SAE

*A Rapid Wave Action Simulation Technique for Intake Manifold Design,* R.J. Pearson, D.E. Winterbone, 1990, 900676, SAE

*Intake-Generated Swirl and Tumble Motions in a 4-Valve Engine with Various Intake Configurations—Flow Visualization and Particle Tracking Velocimetry,* 1990, Bahram Khalighi, 900059, SAE

*Measuring Absolute-Cylinder Pressure and Pressure Drop Across Intake Valves of Firing Engine,* Paulius Puzinauskas, Joseph C. Eves, Nohr Tilman, 1994, 941881, SAE

### Research and Technical Reports

*Flow of Fluids Through Valves, Fittings, and Pipe,* Crane Technical Paper 410, 1957, Crane Company

*Wave Dynamics, Helmholz Resonance, and the Ram Effect in Tuned Intake Systems,* James Sinnamon, Andrew Randolph, Paulius Puzinauskas, 1987, EN-416, General Motors Research Laboratories

*Engine Airflow Development Guide for Combustion Air Handling Devices,* K.D. Sperry, 1989, Chevrolet-Pontiac-Canada Group, Engineering Center

*Motorsports Standard Atmosphere and Weather Correction Methods,* Patrick Hale, 2008, Drag Racing Pro

### HPBooks References:

*Holley Carburetors, Mike Urich & Bill Fisher,* 1987, HPBooks, ISBN 978-1-55788-052-9

*Rochester Carburetors,* Doug Roe, 1981, HPBooks, ISBN 978-0-89586-301-0

*Weber Carburetors,* Pat Braden, 1988, HPBooks, ISBN 978-0-89586-377-5

*Rebuild & Powertune Carter/Edelbrock Carburetors,* Larry Shepard, 2010, HPBooks, ISBN 978-1-55788-555-5

# ABOUT THE AUTHOR

Harold Bettes has been a mechanical engineer for more than 40 years, working in all forms of motorsports, and in the design, development, manufacturing and sales of test equipment and components for high-performance applications. His designs have appeared in special equipment for the motorsports industry, as well as petroleum and aviation industries.

Harold has presented countless seminars on engine airflow and testing for various schools, colleges and OEM groups and the motorsports industry. He has been a guest speaker and panel participant for the American Society of Mechanical Engineers and the Society of Automotive Engineers. He has personally written or contributed to many technical articles for magazines, trade publications and books, including *Dyno Testing and Tuning* with Bill Hancock. He has served on the industrial/academic advisory board (engineering) for the Colorado School of Mines, been recognized with awards by *Car Craft* magazine (1980), ASME (1990) and Standard 1320 Drag Racing Group (2000) and is the recipient of a lifetime achievement award from the Advanced Engineering Technology Conference (2005).

Over the years, Harold has competed with go-karts, motorcycles, boats, airplanes and race cars. He still holds various competition licenses and an unblemished private pilot certificate.

# HPBooks

## GENERAL MOTORS
Big-Block Chevy Engine Buildups: 978-1-55788-484-8/HP1484
Big-Block Chevy Performance: 978-1-55788-216-5/HP1216
Camaro Performance Handbook: 978-1-55788-057-4/HP1057
Camaro Restoration Handbook ('61–'81): 978-0-89586-375-1/HP758
Chevelle/El Camino Handbook: 978-1-55788-428-2/HP1428
Chevy LS1/LS6 Performance: 978-1-55788-407-7/HP1407
The Classic Chevy Truck Handbook: 978-1-55788-534-0/HP1534
How to Rebuild Big-Block Chevy Engines: 978-0-89586-175-7/HP755
How to Rebuild Big-Block Chevy Engines, 1991–2000: 978-1-55788-550-0/HP1550
How to Rebuild Small-Block Chevy LT-1/LT-4 Engines: 978-1-55788-393-3/HP1393
How to Rebuild Your Small-Block Chevy: 978-1-55788-029-1/HP1029
Powerglide Transmission Handbook: 978-1-55788-355-1/HP1355
Small-Block Chevy Engine Buildups: 978-1-55788-400-8/HP1400
Turbo Hydra-Matic 350 Handbook: 978-0-89586-051-4/HP511

## FORD
Ford Engine Buildups: 978-1-55788-531-9/HP1531
Ford Windsor Small-Block Performance: 978-1-55788-558-6/HP1558
How to Build Small-Block Ford Racing Engines: 978-1-55788-536-2/HP1536
How to Rebuild Big-Block Ford Engines: 978-0-89586-070-5/HP708
How to Rebuild Ford V-8 Engines: 978-0-89586-036-1/HP36
How to Rebuild Small-Block Ford Engines: 978-0-912656-89-2/HP89
Mustang Restoration Handbook: 978-0-89586-402-4/HP029

## MOPAR
Big-Block Mopar Performance: 978-1-55788-302-5/HP1302
How to Hot Rod Small-Block Mopar Engine, Revised: 978-1-55788-405-3/HP1405
How to Modify Your Jeep Chassis and Suspension For Off-Road: 978-1-55788-424-4/HP1424
How to Modify Your Mopar Magnum V8: 978-1-55788-473-2/HP1473
How to Rebuild and Modify Chrysler 426 Hemi Engines: 978-1-55788-525-8/HP1525
How to Rebuild Big-Block Mopar Engines: 978-1-55788-190-8/HP1190
How to Rebuild Small-Block Mopar Engines: 978-0-89586-128-5/HP83
How to Rebuild Your Mopar Magnum V8: 978-1-55788-431-5/HP1431
The Mopar Six-Pack Engine Handbook: 978-1-55788-528-9/HP1528
Torqueflite A-727 Transmission Handbook: 978-1-55788-399-5/HP1399

## IMPORTS
Baja Bugs & Buggies: 978-0-89586-186-3/HP60
Honda/Acura Engine Performance: 978-1-55788-384-1/HP1384
How to Build Performance Nissan Sport Compacts, 1991–2006: 978-1-55788-541-8/HP1541
How to Hot Rod VW Engines: 978-0-91265-603-8/HP034
How to Rebuild Your VW Air-Cooled Engine: 978-0-89586-225-9/HP1225
Mitsubishi & Diamond Star Performance Tuning: 978-1-55788-496-1/HP1496
Porsche 911 Performance: 978-1-55788-489-3/HP1489
Street Rotary: 978-1-55788-549-4/HP1549
Toyota MR2 Performance: 978-155788-553-1/HP1553
Xtreme Honda B-Series Engines: 978-1-55788-552-4/HP1552

## HANDBOOKS
Auto Electrical Handbook: 978-0-89586-238-9/HP387
Auto Math Handbook: 978-1-55788-020-8/HP1020
Auto Upholstery & Interiors: 978-1-55788-265-3/HP1265
Custom Auto Wiring & Electrical: 978-1-55788-545-6/HP1545
Engine Builder's Handbook: 978-1-55788-245-5/HP1245
Engine Cooling Systems: 978-1-55788-425-1/HP1425
Fiberglass & Other Composite Materials: 978-1-55788-498-5/HP1498
High Performance Fasteners & Plumbing: 978-1-55788-523-4/HP1523
Metal Fabricator's Handbook: 978-0-89586-870-1/HP709
Paint & Body Handbook: 978-1-55788-082-6/HP1082
Practical Auto & Truck Restoration: 978-155788-547-0/HP1547
Pro Paint & Body: 978-1-55788-394-0/HP1394
Sheet Metal Handbook: 978-0-89586-757-5/HP575
Welder's Handbook, Revised: 978-1-55788-513-5

## INDUCTION
Engine Airflow, 978-155788-537-1/HP1537
Holley 4150 & 4160 Carburetor Handbook: 978-0-89586-047-7/HP473
Holley Carbs, Manifolds & F.I.: 978-1-55788-052-9/HP1052
Rebuild & Powertune Carter/Edelbrock Carburetors: 978-155788-555-5/HP1555
Rochester Carburetors: 978-0-89586-301-0/HP014
Performance Fuel Injection Systems: 978-1-55788-557-9/HP1557
Turbochargers: 978-0-89586-135-1/HP49
Street Turbocharging: 978-1-55788-488-6/HP1488
The Engine Airflow Handbook: 978-1-55788-537-1/HP1537
Weber Carburetors: 978-0-89589-377-5/HP774

## RACING & CHASSIS
Advanced Race Car Chassis Technology: 978-1-55788-562-3/HP562
Chassis Engineering: 978-1-55788-055-0/HP1055
*4Wheel & Off-Road*'s Chassis & Suspension: 978-1-55788-406-0/HP1406
How to Make Your Car Handle: 978-1-91265-646-5/HP46
How to Build a Winning Drag Race Chassis & Suspension: The Race Car Chassis: 978-1-55788-540-1/HP1540
The Racing Engine Builder's Handbook: 978-1-55788-492-3/HP1492

## STREET RODS
*Street Rodder* magazine's Chassis & Suspension Handbook: 978-1-55788-346-9/HP1346
Street Rodder's Handbook, Revised: 978-1-55788-409-1/HP1409

### ORDER YOUR COPY TODAY!
All books are available from online bookstores (www.amazon.com and www.barnesandnoble.com) and auto parts stores (www.summitracing.com or www.jegs.com). Or order direct from HPBooks at www.penguin.com/hpauto. Many titles are available in downloadable eBook formats.